Praise for

Soil Sisters

I wish all my neighbors were SOIL SISTERS — better farmers, better food, better fellowship — what's not to love? Appreciating and leveraging women's distinctives, *Soil Sisters* empowers and guides feminine farmers to full expression. A truly wonderful guidebook.

— Joel Salatin, Polyface Farm

Reading these pages feels more like sitting across from Lisa and friends at a kitchen table, where you quickly come to realize that every female farmer was once an apprehensive beginner and that you're not alone — you are among sisters.

— MaryJane Butters, *MaryJanes Farm*, from the Foreword

Lisa Kivirist's *Soil Sisters* is a book with perfect timing. The past seven hundred years of settler colonialism have worked to crush women's knowledge of farming and agriculture. But knowledge is hard to kill. It lives in first nations, in urban gardens and, nurtured by the soil of her farm, on the pages of Lisa's book. With more and more young women looking for a future on the farm, this book couldn't be more timely or useful. Buy it along with your first shovel and hat.

— Raj Patel, author, *Stuffed and Starved: The Hidden Battle for the World Food System*

This isn't a "how-to-farm" book; there are plenty of books out there on raising sheep or growing garlic. Instead, *Soil Sisters* provides something much more valuable — support and encouragement and concrete tips for those women courageous enough to step into a new life as a farmer. Kivirist brings together a chorus of voices so powerful that it's like being surrounded by women neighbors willing to lend a hand. Their voices — and what they've learned — get me excited about farming all over again.

— Catherine Friend, author, *Hit by a Farm* and *Sheepish: Two Women, Fifty Sheep, and Enough Wool to Save the Planet*

This isn't your grandma's view of farming anymore. *Soil Sisters* sheds light on the new wave of women farmers. From growers to entrepreneurs, from the fields and inside kitchens; this must-have book has anecdotes, advice, tips and stories for women who want to start a business, learn a craft, or dare to dream big. If that's you, this book is for you.

— Karen Washington, Founder, Black Urban Growers;
and farmer & co-owner, Rise&Root Farm

All over the globe, women farmers and gardeners are leading the way to healthier lives and landscapes. *Soil Sisters* celebrates the critical role they play in putting good food on the world's table while offering a hearty serving of information and inspiration for the next generation of female growers.

— Roger Doiron, Founding Director, Kitchen Gardeners International

This book is just what's needed to inspire women new to farming, and it's a great read for those who have been farming for years. The book is full of real life stories, and Lisa writes from the heart about what she knows best: the inspiring spirit of the women farming movement. With authenticity and seasoned experience, *Soil Sisters* provides the much-needed road map for women with organic farming dreams today.

— Faye Jones, Executive Director,
Midwest Organic and Sustainable Education Service (MOSES)

This toolkit for women farmers is a must-read for any woman currently involved in or looking to get involved in agriculture. Women have always played an imperative role in agriculture, nurturing the land and the livestock, but have had minimal resources built to fit their needs and their specific roles. With *Soil Sisters*, women farmers and ranchers can learn from each other in a practical and unique way.

— Marji Guyler-Alaniz, Founder & President, FarmHer Inc.

A long overdue tribute to and toolkit for women farmers, *Soil Sisters* is a priceless slice of strength and power for women of the land. I have never felt more like a woman than the day I first dug my hands into the soil, and digging into this book is a joyful and much-needed affirmation of that day.

— Natasha Bowens, author, *The Color of Food: Stories of Race, Resilience and Farming*

As the title suggests, Lisa Kivirist puts two critically important topics together — women and soil — to express the change we are all working so hard for in agriculture today. In the past year, it has been confirmed, the dominant agricultural input, glyphosate, known as Round up, indeed does cause cancer, is building up in our environment and in our bodies. *Soil Sisters*, through stories and reflections, shows solutions: Women in agriculture are instrumental in leading the much needed change. I am inspired as I read the creative approaches, thoughtful management, tireless work ethics. Yes, women are nurturers and much more. Women who are struggling with leadership in all categories — you are not alone! This book represents the "can do" attitude of extraordinary women who are modeling courage and determination. We will make a difference!

— Theresa Marquez, Mission Executive (ME), Organic Valley Family of Farms

At last we have a roadmap for women that want to launch or improve their farm business. *Soil Sisters* uniquely compiles the rich wisdom of successful female farmers and serves it up in an accessible and inspiring way. It's like having a cup of coffee with 100 seasoned women farmers at once and being able to pick their brains about what works and what doesn't.

— Temra Costa, author, *Farmer Jane: Women Changing the Way We Eat*

As Lisa Kivirist reminds us in *Soil Sisters*, husbandry ain't just a job for husbands. Women farmers are at the forefront of creating a more just, sustainable and profitable relationship with the land, and *Soil Sisters* is here to speed the plow.

— Curt Ellis, co-creator, *King Corn* and co-founder, FoodCorps

Women are getting into farming in unprecedented numbers and they are changing the face of American agriculture. The current food scene — artisanal, direct-to-consumer, local food — is flourishing with the energy and creativity of women farmers. As the publisher of *Growing for Market* magazine for the past 25 years, I have met thousands of amazing women who are either full farming partners with their spouses or running their farms single-handedly. They manage to juggle the hard physical work of farming with raising children, running a household, and maintaining social connections. They lead lives of intensity and fulfillment. In *Soil Sisters,* Lisa Kivirist has brought many of these women to life for her readers. She shares their insights and success stories and, in the process, provides inspiration for women everywhere to consider farming as a career and a lifelong passion.

— Lynn Byczynski, editor and publisher, *Growing for Market*

Lisa Kivirist learned to succeed in farming and now she wants you to as well. You're not just another woman dreaming of the rural life, you're a Soil Sister to her. Lisa offers a unique voice to the perspective of women in the agricultural movement while encouraging, inspiring and instructing in a way only she can do.

— Maria Miller, Women in Ag Advocate and former Executive Director, NFU Foundation

Inspiring, practical and powerful, *Soil Sisters* provides the needed catalyst to keep our women in sustainable agriculture movement growing strong. I've known Lisa for years and she exemplifies from the heart what women farmers are all about — heartfelt collaboration — and *Soil Sisters* captures that spirit with every page.

—Denise O'Brian, founder of Women, Food & Agriculture Network (WFAN)

This practical, accessibly-written, and well-researched treasure will teach you everything from how to choose tools and equipment to how to connect with other women farmers. Through stories of other women farmers as well as her own experiences, Kivirist will guide you on how to get access to land, host farm dinners, take care of your body, and take the lead in "cultivating new approaches to farm-based livelihoods." This book should be on the nightstand of every woman farmer, farmer wann-be, and friends of women farmers.

Carolyn Sachs, Professor of Rural Sociology and Women's Studies and co-author of the forthcoming book: *The Rise of Women Farmers and Sustainable Agriculture*, University of Iowa Press

SOIL
Sisters

A TOOLKIT FOR WOMEN FARMERS

▼▼▼▼▼▼▼▼▼▼▼▼▼▼▼▼▼▼

Lisa Kivirist

new society
PUBLISHERS

Cover design by Diane McIntosh.
Cover Images: illustrative elements © iStock
Interior: Barn door © AdobeStock_87418027; Kids blackboard © AdobeStock_62817701;
Book and feather pen © AdobeStock_93465377
Interior Illustrations: Seed pack, garden tools, canning jar © MJ Jessen
Printed in Canada. First printing January 2016.

Paperback ISBN: 978-0-86571-805-0
eISBN: 978-1-55092-602-6

Funded by the Government of Canada	Financé par le gouvernement du Canada	Canada

Inquiries regarding requests to reprint all or part of *Soil Sisters* should be addressed to New Society Publishers at the address below. To order directly from the publishers, please call toll-free (North America) 1-800-567-6772, or order online at www.newsociety.com

Any other inquiries can be directed by mail to:
New Society Publishers
P.O. Box 189, Gabriola Island, BC V0R 1X0, Canada
(250) 247-9737

New Society Publishers' mission is to publish books that contribute in fundamental ways to building an ecologically sustainable and just society, and to do so with the least possible impact on the environment, in a manner that models this vision. We are committed to doing this not just through education, but through action. The interior pages of our bound books are printed on Forest Stewardship Council®-registered acid-free paper that is **100% post-consumer recycled** (100% old growth forest-free), processed chlorine-free, and printed with vegetable-based, low-VOC inks, with covers produced using FSC®-registered stock. New Society also works to reduce its carbon footprint, and purchases carbon offsets based on an annual audit to ensure a carbon neutral footprint. For further information, or to browse our full list of books and purchase securely, visit our website at: www.newsociety.com

Library and Archives Canada Cataloguing in Publication

Kivirist, Lisa, author
    Soil sisters : a tooklit for women farmers / Lisa Kivirist.
Includes index.
Issued in print and electronic formats.
ISBN 978-0-86571-805-0 (paperback).--ISBN 978-1-55092-602-6 (ebook)
    1. Women in agriculture. 2. Agriculture. 3. Business planning.
4. Businesswomen. 5. Entrepreneurship. I. Title.
HD6077.K59 2016   338.1082   C2015-906831-2
                                                        C2015-906832-0

Contents

PART 4: CULTIVATING QUALITY OF LIFE
   Nurturing your body, mind, and spirit

Dedication

For the inspiring women farmer pioneers
who cultivated the soil that enriches the movement today.

For the next generation of women transforming our food system.

For the many men supporting every seed we plant,
especially John D. Ivanko, who nurtured this book
alongside his wife every page of the way.

Acknowledgments

SOIL SISTERS CAME ABOUT through the wide and wonderful network of farmers and sustainable agriculture activists and supporters I am honored to call my friends. Thank you for everything — from your inspiring stories to those "You go, girl" emails along the last manuscript mile. This list is by far not inclusive, and I'm sure there will be additions to come after we hit print. I am grateful and blessed for how you all have enriched my life:

Ann Adams, Leigh Adcock, Terese Allen, Wendy Allen, Audrey Alwell, Betty Anderson, Smaranda Andrews, Rachel Armstrong, Jamie Baker, Dr. Jenny Barker-Devine, Anna Thomas Bates, Kat Becker, Harriet Behar, MaryAnn Bellazinni, Ali Berlow, Deirdre Birmingham, Chris Blanchard, Natasha Bowens, Zöe Bradbury, Liz Brensinger, Hannah Smith Brubaker, Sarah Broadfoot, Terra Brockman, Traci Bruckner, Melissa Burch, MaryJane Butters, Elena Byrne, Lynn Byczynski, Sarah Calhoun, Dr. Gail Campbell, Sarah Campbell, Patty Cantrell, Gail Carpenter, Cara Carper, Brenda Carus, Luis Carus, Mary Cavanaugh, Katie College, Jamie Collins, Courtney Cowgill, Jordan Champagne, Dawn Combs, Deanne Coon, Temra Costa, Pam Dawling, Katy Dickson, Pam Dixon, Atina Diffley, Regina Dlugokencky, Margaret De Bona, Roger Doiron, Jean Eels, Curt Ellis, Laura Endres, Dela Ends, Danielle Endvick, Deb Eshmeyer, Nina Kahori Fallenbaum, Amy Fenn, Melissa Fery, Catherine Friend, Paula Foreman, Maisie Ganz, Julie Garret, Elizabeth Goreham, Dr. Tammy Gray-Steele, Marji Guyler-Alaniz, Eric Halting, Alissa Hamilton, Pakou Hang, Ann Larkin Hansen, Dr. Rose Hayden-Smith, Shannon Hayes, Willow Hein, Melinda Hemmelgarn, Elizabeth Henderson, Lynn Heuss, Carol Hill, Clare Hintz, Bridget Holcumb, Beth

Holtzman, Jane Jewett, Selket Jewett, Jan Joannides, Liz Johnson, Scottie Jones, Erika Jones, Faye Jones, Susan Jutz, Sonia Kendrick, Katharine Kramer, Darla Krieger, Oriana Kruszewski, Margaret Krome, Sarah Kyrie, Jamie Lamonde, Anna Landmark, Lauren Langworthy, Steph Larsen, Martha Lee, Katie Lipes, Anna Lappe, Sharon Lezberg, Cathy Linn-Thortenson, Sarah Lloyd, Gabriele Marewski, Kriss Marion, Theresa Marquez, Christine Matherson, Jack Matson, Holly Mawby, Barbara Meister, Jen Miller, Maria Miller, Victoria Redhed Miller, Alissa Moore, Lindsey Morris Carpenter, Mark Muller, Lisa Munniksma, Dr. Helene Murray, Jennifer Nelson, Dr. Astrid Newenhouse, Deborah Niemann, Denise O'Brien, Kara O'Connor, Brett Olson, Beth Osmund, Cathy Olyphant, Jody Padgham, Elisa Parker, Raj Patel, Mary Peabody, Katie Peterman, Nirav Peterson, Tracy Potter-Fins, LeAnn Powers, April Prusia, Tom Quinn, Ruth Rabinowitz, Althea Raiford, Marguerite Ramlow, Joy Reavis, Jen Riemer, Susan Roberts, Diane Rogers, Erica Roth, Dr. Carolyn Sachs, Joel Salatin, Maria Sayles, Erin Schneider, Carol Schutte, Kate Seager, Valerie Segrest, Peg Sheaffer, Lisa Shirek, Debra Sloane, Tara Smith, Dorothy Stainbrook, Carly Stephenson, Jane Lori Stern, Hawley Stevens, Bethany Strom, Kate Stout, Angie Sullivan, Angie Tagtow, Rebecca Thistlethwaite, Elizabeth U, Nancy Vail, Paula Vestin, Severine von Tscharner Fleming, Kristi Waits, Karen Washington, Pam Walgren, Relinda Walker, Monica Walch, Christine Welcher, Jesse Wellington, Lauren Wells, Rachel Werner, Inga Witscher, Amiee Witteman, Carla Wright, Jaclyn Wypler, Debby Zygielbaum.

Thanks also go to amazing organizations supporting the women in the sustainable agriculture movement, including Annie's Project, Barnraiser, Center for Integrated Agricultural Systems, Center for Rural Affairs, FairShare CSA Coalition, Edible Communities, Farm Commons, FarmHer, Farm Stay U.S., Green Heron Tools, Greenhorns, Growing for Market, Holm Girls Dairy, Kellogg Fellows Leadership Alliance (KFLA), Land Stewardship Project, MaryJanesFarm, Midwest Organic and Sustainable Education Service, Michael Fields Agricultural Institute, Minnesota Institute for Sustainable Agriculture, University of Minnesota, Monroe Chamber of Commerce and Industry, Mother Earth News, National Farmers Union, National Sustainable Agriculture Coalition, National Young Farmers Coalition, National Women in Agriculture Association, Pennsylvania Women's Agricultural Network, Red Ants Pants, Renewing the Countryside, Sustainable Agriculture Research and Education Program, Seed Savers, United States Department of Agriculture, Women's Agricultural Network, and Wisconsin Farmers Union.

A grateful shout out to the inspiring team at New Society Publishers, talented women who continue to share my vision for instigating change through words. A special thank you to Ingrid Witvoet, E.J. Hurst, Sara Reeves, MJ Jessen, Greg Green and Sue Custance for all your support, and to Audrey Dorsch for adding sparkle to the manuscript. Once again, Diane McIntosh nailed the perfect cover.

Thank you to my parents, Valdek and Aelita Kivirist, and my mother-in-law, Susan Ivanko, for gifting our family with a love of good food.

Appreciation to Liam for showering his mom with writing and tech support every step of the way and for racking up — with a smile — more visits to women-run farms than probably any 14-year-old male in history. Deep gratitude to my husband, John D. Ivanko. From his nurturing of the first idea seeds of this project to reading each word and kindly suggesting I throw out that first draft, the pages of *Soil Sisters* and every day of my life turn out sweeter with his touch.

Foreword

LISA HAD ME WITH THE FIRST WORD of this book's title, *Soil*. But she really had me with the second word, *Sisters*. For thousands of female farmers, soil is our life's passion. It's what gets us up and into our Carhartts every morning. But to do so in collaboration with like-minded women is the ultimate in harvests.

MaryJane Butters of MaryJanesFarm.

Soil represents more than just a spot to sow a seed or something you shake off your boots at the end of the day. We think of it as rich, dark, nutrient-dense organic matter that we nurture through compost and care. Cultivated with pride, the ground beneath our feet gives us the opportunity to bring spinach to life in the early spring. That same patch of ground eventually becomes a pumpkin patch in the fall, through which kids run and squeal in delight. Cows moo, gifting us milk and manure and our free-range chickens give us eggs as they poop their way to improving the health of our soil. It's all connected.

Another word tattooed on my heart is Sisters. It's the key ingredient that has kept my farm running for 30 years, first raising organic vegetables for our local farmers' market, then diversifying into my *MaryJanesFarm* magazine, books, farm school, farmstay bed and breakfast, and more. My name may be the one on my agricultural operation, but I'm the first to say nothing here would be what it is today without the inspiring support of kindred spirits committed to collaboration.

Our Farmgirl Sisterhood program epitomizes exactly that support system, with close to 7,000 dues-paying members (Lisa is Sister #5) who gather both online and in person to share knowledge and learn together — everything from gardening to going green. A grown-up girl's 4-H, we embrace four different concepts that start with H: home, hearth, handiwork, and hogs (or cow or chickens or any of the other cute critters that free-range our fields). From Boise to Boston, from small towns to urban centers, women come together to connect with and celebrate "the farmgirl in all of us."

It's this collaborative strength, this sisterhood bond that fuels the fact that women are one of the fastest growing groups of people buying and operating small farms. While the number of American farms continues to decline, the number of farms run by women has doubled, adding up to about 30 percent of farms in the U.S. Female farmers launching agricultural start-ups favor smaller, more diverse operations with an unparalleled tendency toward sustainable farming practices.

That sharing spirit, rooted in friendship, comes across in every page of this book as Lisa delivers stories, advice, and inspiration from over a hundred women seasoned in various aspects of sustainable agriculture. This isn't some highbrow, academic gathering of specialists. These women are Lisa's own "sisters," her personal farming family who generously and authentically share what's growing on their side of the pasture fence. Reading these pages feels more like sitting across from Lisa and friends at a kitchen table, where you quickly come to realize that every female farmer was once an apprehensive beginner and that you're not alone — you are among sisters.

More than just a place of business, our farms provide us with a blank canvas on which to express ourselves. Here's a fun female farmer fact you won't find in an agricultural census: We love what we do and have fun doing it. Playful and innovative, we use our farmsteads as a way to share what makes our hearts sing. You'll meet Gabriele Marewski of Paradise Farms in Florida, who paints her outbuilding doors pink and wears a skirt when working in the field simply because she can. Wallpaper the inside of your chicken coop? Been there, done that.

With incredible ingenuity, we're feeding our creative urges through our farms. When we raise lambs for sale, we give lessons on spinning and knitting. When we grow the ingredients for salsa, we hold Latin dance clinics. When we promote the sale of our crops as ingredients perfect for pizza, we plant our vegetables in wedges that form a grand circle when seen from afar. When we grow 60 different sunflowers, we take photos for a line of sunny greeting cards. When we decide to convert land into native prairie again, we turn it into a business that sells and promotes endangered native plants. When we open up our hearts, we open up our homes and turn them into bed-and-breakfast sanctuaries where our guests are fed the best food on Earth.

MaryJane's three granddaughters at their farmers' market booth, selling organic heirloom "petite" tomatoes.

The message: Be yourself and bring your quirky, creative moxie to everything you do.

Just like Mother Nature, women growers build farms and livelihoods rooted in diversity. We don't plant just one type of tomato; our fields erupt in an array of colors and flavors, from Green Zebra to the yellow Lemon Drop. Likewise, we ourselves come from a diversity of backgrounds, communities, and cultures. We believe food can build bridges and heal. Dropping off your fresh zucchini muffins transforms that crabby curmudgeon bachelor farmer next door into an ally (warning: this might take multiple deliveries!); and sharing our abundance with sisters in need by offering reduced pricing on our CSA vegetable boxes helps them feed their families healthy food.

We've come a long way as female farmers, especially in the last couple of decades. When I applied for status as Idaho's first organic food manufacturer in the early 1990s, I was turned down. "We don't want you putting the word 'organic' on your food," said a politician with strong ties to big business. "When you use the word 'organic', you charge more for it, making it seem like your food is better than ours." Years ago, a banker asked me if there were any men in my life who would be willing to co-sign the agricultural operating loan I had applied for. Women were both rare and not respected back when I started farming in 1986. While things are indeed different today, we need to understand our roots and histories in order to better take on the future. Soil Sisters provides the road map for just that.

We still have a long way to go. Women add up to a relatively small piece of the national agricultural pie. Even though we account for roughly 30 percent of all farmers, we control only 7 percent of American farmland that accounts for 3 percent of agricultural sales. There isn't a Farm Bill program that specifically addresses the needs and opportunities of female farmers.

I may be writing this from my farmstead in the outback of Idaho, looking out upon acres of emptiness without a person in sight, yet I never feel alone. I know my Soil Sisters are with me from the moment my rooster crows until I step back outside one last time to check on my Jersey cows under a starry night sky. Remember, you have me, Lisa, the over 100 women in this book, and the thousands of women throughout the country who not only understand, but embrace your tractor fantasies. We're more than farmers, we're family.

MaryJane Butters is an Idaho organic farmer, book author, and the editor of *MaryJanesFarm*, a magazine with 150,000 subscribers. She also operates a small dairy, sells 60 different prepared foods packaged at her farm, and owns a historic flour mill ... an all around entre-manure!

Preface:
Land — A Love Story

IN 1976, AT AGE NINE and growing up in the Chicago suburbs, I painted a picture of a barn and silo in art class. It came out so beautiful that my dad insisted on custom framing the creation and hanging it on the wall. That same year I pleaded with my mom to sew me a Laura Ingalls-inspired Halloween costume — complete with sunbonnet. I rocked the *Little House on the Prairie* fashion sidewalk runway that season.

Left: *Lisa Kivirist's farm dream, circa third grade art class.*

Right: *Lisa Kivirist channeling her inner Laura Ingalls, Halloween 1976.*

Dutch oven bread
baking at Inn
Serendipity.
John D. Ivanko

Flash forward 40 years. I now own that barn, and Ma Ingalls would groove on my bread-baking bonanzas — especially in my Lodge Dutch oven. She'd want to sample my rows of jars of homegrown pickles and sauerkraut.

So, you could blame my parents for unknowingly encouraging me to become a female farmer. It took several decades and detours to get from the barn artwork on the wall to a barn on my farm, let alone more than 40 vegetable varietals growing in my farm fields. Other than my enthrallment with *Little House on the Prairie*, the word "farmer" had never found its way into my career radar.

Though I grew up in a food-loving family that embraced gathering around the table over a meal, a separation existed between the turkey on the platter and the farm where it was raised. I was the classic suburban-raised kid of the 1970s, an era in which processed foods started popping onto our family tables and "convenience" was lauded as progress.

Post-college and armed with a communications degree from Northwestern University, I followed the expected career norm, marching into a corporate cubicle and working at an advertising agency. My first client: Hallmark. Parents were pleased, of course. Mom and Dad had important visions for what their only daughter should be. A dirt farmer was not one of them. On the outside, everything appeared normal: long hours under fluorescent lights, pre-approved credit card offers piling up in the mailbox, and take-out food in Styrofoam containers quickly consumed during 15-minute lunch breaks. But in reality, I was well on my way to a premature mid-life crisis at age 25. Paychecks and promotions didn't fulfill me in the way they seemed to satisfy most everyone else. Which left me challenged with the question: What will?

My self-created therapy in dealing with this lack of fulfillment consisted of weekend escapes to Wisconsin. A colleague in the cubicle world, John Ivanko, started a social group for us disgruntled young professionals called the Outing Club, organizing informal camping or biking trips across the border into the rolling green hills of America's dairyland. Now, realize I had never been camping until this point; our family vacations always involved air conditioning and indoor plumbing. But something about John's hand-drawn Outing Club flyer in the office break room caught my eye. Or maybe John caught my eye first, as I quickly developed a crush on this guy who could pass for Harrison Ford's double. Whatever my reasons — an antidote for corporate malaise, true love, or my Han Solo fantasies — I enthusiastically signed up for the next camping weekend, a trip to the New Glarus Woods State Park.

Life's detours can entertain us and, if we let them, fundamentally sweeten our lives. The first time I met the guy behind that flyer, I knew I was destined to ride through life's journey with him. John, you see, was my soul mate and eventually became my husband. We shared a yearning for a different life path, but we didn't have a clear direction back then on what or where that would be. So as an immediate fix, we kept organizing those escapist Outing Club weekend trips. Getting physically and mentally out of the cubicle recharged us like no amount of double espressos ever could.

I never thought of myself as a farm girl or someone who wanted to live in the middle of nowhere. Heck, I never even slept in a tent until that infamous first camping trip. While my yuppie peers were talking promotions and corner offices, I found myself counting the hours till the next trip out of the city, when I could linger on those rural back roads, sip that 79-cent cup of coffee at the corner café, and see the Milky Way. It was the first time I smelled a freshly manured field and played with a barn kitten.

On some trips we upgraded from camping and stayed at bed and breakfasts or farm stays, especially if we were biking all day and craved a hot shower and real mattress. That was the case one sultry summer night when John and I and a half dozen other Outing Club friends arrived at Waarvick's Century Farm Bed and Breakfast, located near the Elroy-to-Sparta bicycle trail. The convivial hostess greeted us with homemade plum liquor, and we toasted the bounty of summer as the fireflies twinkled in the twilight. The next morning, I woke at sunrise with energy that outpaced anything I felt during the work week. I followed my nose to coffee brewing and zucchini muffins baking with a waft of cinnamon. I was given the task of harvesting rhubarb from the garden; thanks to her tour the night before, I knew what I was looking for. I wandered barefoot to the garden, toes wet with dawn dew. Pulling that pink rhubarb as the day broke and the doves cooed in the barn, I felt at happy peace for the first time in a long time.

Other trips introduced us to many more eccentric and creative farmer and rural types who had in common a love of the land and sharing it with others. Some were five-star quirky, like the host at Lonesome Jake's, who arrived to cook breakfast in head-to-toe cowboy regalia and holding a bottle of champagne. Maybe it was the coyotes' howl at night, the fresh scent of a field of cut alfalfa, or that 50-pound pumpkin I insisted we bring home, but what started as weekend flings over the border snowballed into a full-throttle love affair with farm life. Call it my homecoming.

Hanging out and dreaming together back at urban coffee houses during the week, John and I took our relationship with each other and our love affair with the country life to the next level. Together we plotted our move to a farm. We paid off student loans, saved as much cash as we could, and planned the transition from the corporate maze to living on a country road.

For the first time in my adult life, something bigger than myself drew me in and pushed me forward. I studied real estate farm listings and tutored myself on septics and wells. This, of course, was before the online information super

Wisconsin's bucolic rural countryside.
JOHN D. IVANKO

highway. The local realtor who kindly adopted us would fax black-and-white farm MLS listings. My mind would fill in the picture with abundant vegetables, rows of resplendent flowers, and fragrant herbs. I imagined chickens running around and maybe, one day, a kid nestled up in a tree. My dream livelihood was more than a paycheck and company retirement plan. I wanted freedom to create and be myself on a farm.

No surprise, becoming a farm girl freaked out my parents. Much blame for these crazy ideas was laid on John. After all, he's the one who invited me to go camping in the first place. If only I had stayed in the mall, shopping at Target and Old Navy like normal, manicured women. To add fuel to their fire of irritation, John went back to graduate school for a degree in, of all things, leisure studies. Try explaining that topic to your future in-laws! The man to marry their only daughter studies what people do when they are not working. WTF? Some daughters go through that rebellious phase at age 16. Mine came at 25, when the word "farmer" entered my career vernacular.

But this was not an escapist farm fantasy game for me. I didn't want FarmVille. I demanded the real thing. My heart and mind synced with visions of running a farm business serving garden-fresh meals to bed-and-breakfast guests and feeling the late afternoon sun on my face as I weeded rows of fragrant basil and tomatoes. Just thinking about harvesting strawberries charged me with self-confidence in a way I never encountered crafting a marketing plan for Mother's Day cards at Hallmark. The farm vision gifted me with more than a fresh start; it enabled me to embrace the inner entrepreneur I never knew I had.

November of 1996 racked up to be a month of significance: John and I married; we moved to the five acres of Wisconsin ground we call Inn Serendipity Farm and Bed & Breakfast; we bid farewell to cubicles, high-heeled pumps, and suits and ties. A century-old farmstead, this property epitomized to us what Wendell Berry so eloquently proclaimed as having "faith in two inches of humus that will build under the trees every thousand years" and our willingness to "ask the questions that have no answers." Or at least questions that don't have only one correct answer.

This would be the land on which we'd build a life and livelihood over the decades, in a rural county with twice as many cows as people. At age 30 I was finished with fast food, and I broke the earn-spend treadmill I had been on. I share my journey writing in the second-floor office in the farmhouse as the sun rises and life along County Road P wakens.

Inn Serendipity farmhouse, Browntown, Wisconsin. JOHN D. IVANKO

Today, we have an eclectic patchwork of business ventures, writing, and contract work. They keep both the farm fields blooming and keyboard clicking. I have the freedom to say yes to projects and opportunities I believe in. Gone is the commute to an off-farm job for a paycheck. Raising my own food, serving meals at the B&B, and selling produce to guests collectively make positive deposits into that account of life that give true meaning, the things that hold lasting value. Writing — whether through blog posts, magazine articles, or books — allows me to share inspiring stories and move people to personal action. It's done while peering out the window of my farmhouse office at the sprawling birch tree complete with a robin nest. Priceless.

The learning curve ran steep those first years. I lugged my trusted *Rodale Organic Gardening Guide* to the field to look up which direction to plant the

Map of Inn Serendipity.

seed potatoes (eyes pointing up) or determine if I had tomato hornworm damage on my plants. The book's pages became soil stained as my knowledge base
deepened and the harvest became more prolific.

Besides the rain essential to my crops, I discovered I needed connections to
the waves upon waves of women farmers who shared similar stories, fates, or
epiphanies. Every opportunity to connect with a fellow female farmer strength-
ened my commitment to agriculture and refueled confidence and passion to
get back in the soil. Whether it was a woman I serendipitously sat next to at a
farming conference or a conversation struck up with a woman also loaded with
food preservation books at the library check-out line, I left inspired. Together,
we're more powerful than isolated trees in a forest. There's nothing like knowing you have a sister with a sun hat and calloused hands who both has your
back and challenges you to push yourself.

Take Faye Jones, for example. As executive director of the Midwest
Organic and Sustainable Education Service (MOSES), the hub of organics in

the Midwest, Jones called out of the blue in 2008 and asked if I'd be interested in developing a women-farmer-training program. At the time, I had been increasingly writing on women farmers, telling the stories behind the wind-burned faces sporting Carhartts. Thousands of inspiring women are committed to not just growing fresh, healthy food but transforming our food system in the process.

But it took my soil sister Jones to ignite the spark to get me to think bigger. She saw in me potential I didn't realize I had. Jones created the opportunity and the platform to turn my work and passion to a vocation and — *voilà* — the MOSES Rural Women's Project was born. I've facilitated over 150 trainings, touching thousands of aspiring and seasoned women farmers and including everything from on-farm workshops, such as "In Her Boots: Sustainable Agriculture For Women, By Women," to Organic University conference intensive sessions. The feedback from attendees centered around one point: Please do more of all of this!

Soil Sisters provides that "more" in a take-home version, to reach a broader population of women who might not yet be connected with the farmer-training scene but share a dream like mine of harvesting strawberries, serving

"In Her Boots" workshop with farmer Barb Kraus of Canoe Creek Produce, Decorah, Iowa.
JOHN D. IVANKO

breakfast for farm-stay guests, and hosting potlucks. This book compiles the learning and insights from our trainings in the field in paperback form. It's the book I wished I had back when I started.

Most of the female farmer friends you'll meet through these pages earn much more of their income directly from vegetables they grow, animals they raise, or value-added products produced than I do. There is a patchwork of paths we women take to work toward changing our food system — different squares of the same quilt. Some farm; some cook; some teach; I write. Collaboratively we work together and support each other in the roles we play.

I'm grateful to and inspired by Jones and the women leading change in our food system I've met along the way. This personal connection and appreciation planted the seed for *Soil Sisters:* We have so much to learn and inspiration to draw from each other as women.

The two key words in the title exemplify this spirit:

Soil

Anyone who has planted a seed knows that soil is life. Healthy soil combines various elements, minerals, organic matter, liquids, and other organisms that together support plant life and all terrestrial life, right up the food chain. There are more soil microorganisms in a teaspoon of healthy soil than there are people on Earth. According to the USDA's Natural Resources Conservation Service (NRCS), "Millions of species and billions of organisms — bacteria, algae, microscopic insects, earthworms, beetles, ants, mites, fungi and more — represent the greatest concentration of biomass anywhere on Earth."

Soil can be improved through care, feeding it healthy nutrients just like you feed your own body. My farm's soil needed nurturing big time when we first arrived. Decades of former owners spraying chemical pesticides and herbicides had taken their toll. I've been on a mission to restore this soil to life through adding compost, manure, and "green manure" cover crops of hairy vetch, sweet peas, and oats. Over time, this loam took on such a rich, moist dark color that after it rains I do a double take and think it's brownie batter, ready to sample. Worms wiggle freely and life reigns abundant in the humus. Yields are up too.

As I worked to enrich the health of the soil in the fields, I found myself growing personally as well. The longer I lived in a place of natural and seasonal beauty, from the crisp night air in January to those first ripe, red, tender strawberries in June, the more I connected with my true self and calling. Cucumbers and cabbage need that rich loam in the field to grow. We women thrive when we're surrounded by a healthy situation in which we can express our natural talents. We grow to our authentic selves when we're in a fertile, happy place. *Soil Sisters* poured from my overflowing mug of passion to help support other women, particularly those who might not yet find themselves planted in healthy soil and rooted in an environment of support. This book aims to help you to find your field, to connect you to your dream livelihood by sharing the experiences and ideas of other women further up the road but on the same path.

Sisters

Kudos if you have a biological sister with whom you're on the same page and tightly connected. As an only child, I'm often asked, "Weren't you lonely growing up?" No, I reply, despite coming from a small family with barely any female relatives aside from my mom. Motivated by the lack of sisters or female blood relations, I sought out women outside my family ties, creating my own "sisterhood" of kindred spirits.

It's a powerful force to know you're not alone and someone has your back. You feel empowered when you can expand this definition of "sister" to other women who support you and will catch you when you fall, female friends who will dust you off, give you a hug, and set you back on the tractor.

Introduction: About This Book

I WROTE *SOIL SISTERS: A TOOLKIT FOR WOMEN FARMERS* as a platform, a springboard, to keep our women farmer movement propelled forward, mobilized, and collaborative. Just for the record, this book is not motivated by a lack of general beginning information or resources on farming. As a matter of fact, there hasn't been a better time for beginning farmers of any gender, experience level, interest, and background to tap into the wealth of resources, many available free or at low cost (such as self-guided online portals through the USDA and Extension or formalized curriculum at local community colleges). There is a slew of hands-on, on-farm apprenticeships. You can work at a variety of farms through World Wide Opportunities on Organic Farms (WWOOF) or take a free organic transition course online with the Rodale Institute. Connect the resource dots and fill in knowledge gaps to meet the needs you might have. This book will touch on many.

What's missing from this resource mix, however, is the female farmer voice. Women bring to the training table perspectives, issues, and needs that either get diluted or are absent entirely from current farmer training. Plus, their inspiring stories and experiences pack a practical punch. Each of us comes with our own "operator's manual" that we've self-developed over time that, thanks to the collective nature of this movement, we're eager to share. Like the diversity and abundance found in nature, we represent myriad manifestations of transforming our food system.

This book will connect you with over 50 of these fellow "soil sisters," inspiring women from a range of backgrounds and perspectives who offer pragmatic nuggets on crafting a livelihood in agriculture. Though the bulk of the book focuses on the farmers themselves, it also includes stories from women who are launching various ventures inspired by and related to agriculture and women farmers, although they may not be the ones literally planting the seeds in the field.

The intent of *Soil Sisters*, as a toolkit, is to connect you with resources, tips, and new visions that might help guide your journey and achieve the success you've dreamed about, whether you rank as a seasoned grower or this is the first agricultural publication you've picked up

Be open as you read these pages; there may be concepts you haven't yet thought about. Embrace the possibilities and widen your outlook. Trust me, I didn't have a vision for running my farm on renewable energy when my husband and I first arrived. When the laundry kept blowing off the line, it sparked the idea that perhaps this site was windy enough for a wind turbine to power our farm. A toolkit is about empowerment, the ability to feel confident and prepared to handle any situation. More than a wrench used to tighten the bolts on a moldboard plow, this toolkit gives you the practical means needed to cultivate your dream farm livelihood.

This book complements and builds upon the first book John and I co-authored: *Rural Renaissance: Renewing the Quest for the Good Life. Rural Renaissance*

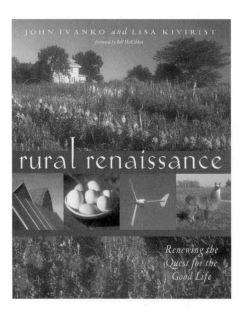

narrates the story of our move from Chicago to our five acres in Wisconsin, including starter resources to get you going in anything from organic growing to renewable energy systems and green design. *Soil Sisters* also partners well with our second book, *ECOpreneuring: Putting Purpose and the Planet before Profits*. *ECOpreneuring* details the business how-to behind creating a livelihood reflecting your sustainability values. The nicest compliment on the book was that it taught the reader everything never covered in business school. It even includes a copy of our tax records so you can better understand how we make our numbers work, like our use of depreciation, investments in renewable energy, and the documentation needed to support such items.

Soil Sisters builds on these themes but viewed through a female lens. Women power the pages of this book: All the quotes and stories, right down to recommended books, come from women who own farms or work in the food sector.

Explore the following *Soil Sisters* themes, organized by book sections:

Part 1: Understanding Our Roots

Women farmers come from a complicated history of growing food for centuries, rooted in discrimination. We bring different experiences and perspectives to our farm visions. By understanding our past and how we fit in to today, we empower ourselves to cultivate a healthy, successful future.

Part 2: Gleaning Knowledge

Develop an understanding of the organic market and how to amp up your knowledge base to succeed in your farm business.

Part 3: Plowing Ahead

Explore various farm business ideas — such as field production, farm-to-table meals, and value-added products — that diversify and thereby strengthen your venture. Explore business planning, marketing, tools, and ergonomics through a female lens.

Part 4: Cultivating Quality of Life

Learn from fellow soil sisters about how to plow your own way. From finding our tribe and dealing with discrimination to integrating family and keeping life in balance, we grow stronger with the support of other women. On a final note, we'll look at catalyzing women's leadership and the role we can play in transforming our food system.

Look for these identifying icons throughout the book as you harvest tips and ideas:

Start-up Stories: "How she sows it" "How I got here" stories from seasoned and successful women

Tool Shed
Practical tips and resource lists

Idea Seeds
A diversity of inspirational quotes from farmers and others related to agriculture and women's issues

Tip Jar
A collection of advice nuggets from fellow soil sisters to kick off each section

With two decades of farming under my tool belt, I've been through the growing season cycle enough to know that snap peas can go in as soon as the soil can be worked up, but hold off on those green beans until after our mid-May frost date. It's tempting for me to get overly excited and move ahead too fast in the fields, only to lose tender plants to a late frost. Take things slow and steady, I've learned. Quality takes time. Practice. Be observant, noting subtle changes or the appearance of a Colorado potato beetle. Where there is one, there are many.

I especially need to nurture my cukes because as soon as summer rolls in, my parents will be asking, "Where are my cucumbers?" After nearly two decades, my mom and dad finally warmed up to the idea of their daughter, the farmer. They may not fully understand or embrace everything I do in a place my mom still calls "the boonies." But when I bring them that bushel of fresh pickling cucumbers, their faces light up. Farm life and the bounty it produces can bridge many divides. My 90-year-old father loves these cucumbers so much you can even catch him slicing one up for a summer breakfast, sitting in the house where my barn and silo masterpiece still hangs on the wall.

Lisa's son, Liam, with cucumber harvest bounty. John D. Ivanko

PART 1:
UNDERSTANDING OUR ROOTS

Origins and polyculture dreams

- Honor your elders.
- Learn from mistakes.
- Remember where you came from.
- Collect canning jars.
- Continually compost and build the soil.
- Look up and realize the wealth already around you.
- Understand that experience is your best teacher.
- Make friends with uncertainty.
- Invite the senior lady next door over for coffee.
- Share your failures and successes.

Chapter 1

Fertile Ground

TRANSPLANTING USED TO MAKE ME VERY NERVOUS when I first started farming. There's something about gently holding tiny leek seedlings between my two fingers. They look more like a blade of grass with a clump of delicate roots on one end. My maternal instincts kick in. I feel the urge to protect the leeks from the cold of late April. Quick, get them tucked into some rich, loamy soil; then water to prevent those roots from drying out.

Come harvest time six months later, my relationship with those baby leeks evolves to a power struggle. I need a damn potato fork to dig out those buggers. The mature leeks' roots are now so entrenched in the soil it takes serious effort to harvest them. I'm careful, however, to avoid damaging the roots. Leeks with an intact root system store in our root cellar readily for months.

Those leek roots quickly embraced their soil and established a strong hold and spread out to absorb nitrogen, phosphorus, potassium, and other nutrients.

According to Colorado State University Extension, 80 percent of plant disorders begin with root and soil problems. A plant becomes stable through strong roots, safely anchoring the plant and lessening the threat of soil erosion. Roots significantly affect a plant's ability to adapt to different soil types as well as its ability to tolerate stress and change, which increasingly happens as a result of climate change. According to the Academies of Science, representing 80 countries, 95 percent of climate researchers actively publishing climate papers endorse the consensus position that climate change exists. The Environmental Protection Agency (EPA) cites that the Earth's average temperature has risen

by 1.4°F over the past century and is projected to rise another 2 degrees, to 11.5°F, over the next hundred years.

Roots represent our foundation, that grounding from which we grow. Our history plays a leading role in determining our future. Our roots shape our perspectives and outlook in life and our future farming endeavors. Roots also define our relationship and connection to our past. Before catapulting into careers in farming, let's step back and understand from where we came. We're part of a greater past, shared with our soil sisters who have been growing food for centuries. Unfortunately, much of this work has gone unrecognized or underappreciated in most of history. Taken for granted. Anyone traveling to some so-called developing countries may see this social inequality still vividly playing out today.

In this chapter, we pay homage to and learn from our shared history as women in agriculture. I'll examine our personal history, reflecting on our life experiences to this point and considering how we can use them to fertilize our farming future.

Condensed history of women growing food

While female farmers appear to be the new fresh hipsters, step back and place today's trendiness in historical perspective. Our boots have been on the ground growing food for a long, long time. Families, communities, and the world eat

Idea Seed:

"I have never felt more like a woman than the day I dug my hands into the soil for the first time. I'll never forget the feeling of my hands after my first season of farming."

— Natasha Bowens, author of *The Color of Food*

Natasha Bowens volunteering on an urban farm in Brooklyn during her hopping days.

because of women. The paleo diet may be the modern rage, but women were the first food gathers back in those cave people days of yore.

Agriculture involving crop seeding and domestication of plants and animals started popping up on the human history radar around 10,000 years ago, and this role of household food management continued for women as food preservation and preparation techniques evolved from fermenting to baking. For those of you with kids in the house, someone in the past 24 hours probably has wandered through the kitchen inquiring, "What's there to eat, Mom?" From the cave to the contemporary pantry, we're the ones the world turns to when hunger hits.

While this role of food provider and preparer serves a vital role in society, female contribution in this category has historically gone unrecognized. If a tree falls in the woods and it isn't posted on Facebook, did it really happen? Status updates can be misleading. How many seeds have we planted and peas have we shucked? While we know how many hamburgers McDonald's sold last year — 75 per second, adding up to over 225 million — we have no idea how many home-made dinners have been served by women over that same period. In 1994, McDonald's executives announced at the annual owner-operator convention that they would stop counting hamburgers served because the count had surpassed the 99 billion hamburger mark. Franchise operators changed their signs to the catchall "billions and billions served." I'm tempted to hang a little glowing neon on my farm door too: "Woman-grown food served inside. Way beyond billions served, for 10,000 years and counting."

Idea Seed:

"My grandmother's laughter and her love of pie, the farm, and her family are the roots of my work. Living with fullness is honoring your ancestors and allowing their hopes to stay alive through you and your work."

— Nancy Vail, co-founder and co-director of Pie Ranch, Pescadero, California

Invisible work and grass ceilings

Invisible and invincible. Two words close in spelling but worlds apart in meaning. They both narrate the story of the history of women farming and raising food. Until as recently as the late 20th century, women's farm work and accomplishments never received much recognition when considered against the male-dominated paradigm. It's like our tomatoes harvested or eggs sold never existed. The Census of Agriculture, conducted every five years by the National Agricultural Statistics Service (NASS), the statistical branch of the USDA, provides comprehensive agricultural data, which plays a big role in determining our nation's agriculture and food policy. The census started counting women farmers only in 1978, two years after I dressed up as Laura Ingalls for Halloween. Despite this lack of acknowledgment, our boot heels stood firmly entrenched into the soil, fostering a foundation for a legacy that is blooming and — slowly — being recognized today.

The fact that, throughout most of history, women could not legally own property played a major role in gender inequity in agriculture. In 1887, one third of states still provided no protection for a married woman's property. Once a woman personally sold anything grown on the farm, even if she raised it 100 percent herself, it legally transferred to her husband. By the 1900s, nearly every state had laws in place allowing a married woman to own property and sue in court, but practices and customs prevented their widespread adoption and situations were always even more challenging for women of color.

"Well into the twentieth century, farm assets were still in the husband's name," shares Dr. Jenny Barker-Devine, associate professor of history at Illinois College and author of *On Behalf of the Family Farm: Iowa Farm Women's Activism since 1945*. "The IRS classified the wife as a dependent, which made it extremely difficult for a woman to claim property and farm assets of her husband in the event of his death or a divorce." Property rights most likely went to the next of kin, such as brother.

"Understanding our history makes us stronger and able to plan for a better future," explains Dr. Rose Hayden-Smith of the University of California, a leading authority in the history of women growing food. She is the author of *Sowing the Seeds of Victory: American Gardening Programs of World War I*. "Women have long played an integral role in raising food,

Idea Seed:

"We refer to nature as 'Mother Nature,' and to Earth as 'Mother Earth.' As a woman, awareness of our innate wisdom and connection grounds me, gives me confidence to push past limiting beliefs, and helps me blossom in my work. I believe it is time for women to reconnect and raise the energy and consciousness of our planet. Each of us, in our own individual ways, will help heal our Earth."

— Nirav Peterson, founder of Mother Earth Solutions, co-founder of Hendrikus Organics, Issaquah, Washington

from home gardens to the farm fields, yet those contributions have gone largely unrecognized and thereby undervalued."

One such example is the role women played in food security during World War I. Blending patriotism with planting parsnips, the Women's Land Army of America (WLAA) officially organized in 1917 and ultimately recruited nearly 20,000 largely middle-class urban and suburban women during WWI to enter the US agricultural sector and work as wage laborers at farms, dairies, and canneries, often in rural areas where American farmers urgently needed help while the male labor force fought on the front lines.

"The WLAA wasn't all about patriotism. Many of the women involved saw this as an opportunity to earn a wage and advance the idea that women were fully capable of such work," adds Hayden-Smith.

Women's Land Army of America recruitment poster, 1918.

Inspired by similar programs started a few years earlier in the United Kingdom and Canada, the WLAA brought women of differing economic and geographic backgrounds together to till the soil, bridging urban and rural experiences as well as class differences, and challenging contemporary stereotypes relating to women's role and work.

"Thousands of women, including wealthy socialites and debutantes, took an enormous chance by stepping out of their traditional roles to join a home front mobilization effort. By becoming farmers, these women challenged society's understanding of the kind of work that women were capable of," continues Hayden-Smith. "Before we earned the right to vote in 1920 with the passage of the 19th Amendment, we served our nation with our hands in the soil. The WLAA played an important role as part of this larger argument that women deserve full citizenship...."

"The message reversed after WWI when the men returned from war and took their farm jobs back," says Hayden-Smith on why the WLAA didn't continue during peacetime. "When the war ended, it was seen as an act of patriotism to give your job to a returning soldier."

An underlying cause of this lack of representation of women in agriculture stems from women's traditional role as wives. "Historically, a woman was connected to the land through the men in her life, whether it was her father, brother or, most likely, a husband in marriage," adds Barker-Devine. "Women primarily identified as wives and performed work vital to the success of the farm, including child rearing, which provided the farm labor force."

These women kept their family farms afloat through the Depression by selling the produce, eggs, and other items they raised. In 1938, the Farm Security Administration estimated that women's income from poultry and dairy operations provided 40 to 50 percent of the household budget. These numbers don't even include alternative entrepreneurial ventures like selling baked goods, crafts, and homemade textiles. "For Depression era families, women's

Women selling chickens at the farmers' market, Weatherford, Texas, 1938.

earnings made it possible for them to stay on the farm when commodity prices barely paid the property taxes, let alone provided a living," Barker-Devine explains.

Farm work was immensely labor intensive through the 1950s: jobs women readily supported, such as food production, preservation, laundry, sewing clothes, and procuring water. Additionally, farm women in the first half of the 20th century served as the important glue that held communities together. "More than just being a good neighbor and saying hello, farm women provided support for one another through major life events from birth to marriage to death," adds Barker-Devine.

Victory Garden poster during World War II.

Agencies like Extension reinforced this dependent role by emphasizing women's expected household duties in nutrition, cooking, and gardening. "While this kind of institutionalized sexism definitely existed, it's important to remember that Extension just created this programming based on the requests of these women," Barker-Devine explains. "Farm wives simply wanted to be able to do their jobs as best as they could. However, by doing so, these women played a role in buttressing these farm wife stereotypes."

The WLAA revived during WWII, along with the Victory Garden program of growing produce domestically. "Over 40 percent of fruits and vegetables consumed on the American home front during WWII were grown in school, home, community and workplace gardens," adds Hayden-Smith. "This probably adds up to the highest consumption of produce ever in America's history." Items such as

sugar and butter had to be rationed during wartime, but families realized that if they had a garden, they could eat as much as their family needed out of that.

Once again, women's employment in the fields and factories was encouraged only as long as the war effort was on. When WWII ended, federal and civilian policies replaced women workers with men, and we entered an era touting convenience in the home kitchen with media messages pushing Hamburger Helper and Betty Crocker cake mixes. With the GIs coming home and the middle class and car culture growing, the ideal of raising your own food via a farm or home garden fell by the wayside.

Big ag and women's role

Traditional farming and practices went through a transformation after WWII. Enter the age of chemical and industrialized approaches.

"Get big or get out." So went the infamous mantra of Earl Butz, secretary of agriculture under the Nixon and Ford administrations in the mid-20th century. Industrial wheels of change started rolling to heed his call, moving to corn commodities that led to many of the food issues we struggle with today, from high fructose corn syrup driving obesity rates to the proclivity of cheap, processed, fake food. While public awareness of such damage from our industrialized food system grows thanks to increased media awareness and movies like *Food, Inc.*, one element overlooked is the impact on the role of women farmers as big ag obliterated the small-scale, economic vibrancy we had created through the first half of the century, and small poultry, dairy, and vegetable operations disappeared.

On the positive side of the capital- and chemical-intensive agriculture of the 1960s and 1970s, it opened new doors and interests for women on the farm. Women entered roles required of such larger operations, such as handling the accounting and bookkeeping. Reforms to estate tax law started to take place as women began to ask for more protection of their property and finances and demanded

Idea Seed:

"Whomever controls our food controls us. The root of our democratic right to rule ourselves lies in our ability to feed ourselves. We can grow ourselves out of this mess because I believe plows are greater than swords."

— Sonia Kendrick, founding farmer,
Feed Iowa First; army veteran featured in the film,
Terra Firma: A Film About Women, War and Healing

that their names appear on legal documents and that they be recognized as cooperators on the farm.

Farm women provided the leadership for many of these changes. Take Doris Royal, for example. She farmed 240 acres with her husband in Nebraska, land they purchased in 1960 for $72,000 with her husband listed on the title. By 1975, the real estate value had ballooned to $300,000, which Royal calculated would result in $32,000 in real estate taxes that she would owe if she was widowed or divorced, even though their farm income remained the same the whole time. Rallying other farm wives with a petition drive, Royal organized a delegation to carry those signatures to Washington, DC, resulting in both national press and inheritance tax reform. The new legislation identifies and accounts for the work that women put into farms as valid claims to the land.

Crisis, community, and CSAs

As the 1980s rolled into the 1990s, we saw the roots of the women in sustainable agriculture movement emerging. Stemming from different perspectives and motivations — some drawing inspiration from Betty Friedan and the feminist movement and others driven to save the family farm during the farm crisis of the 1980s — women's leadership and innovation in sustainable agriculture unfolded in two key ways:

Community Supported Agriculture (CSA)

Community Supported Agriculture (CSA) started to emerge in the mid-1980s. The idea that customers pay up front for a "share" and then receive weekly produce boxes throughout the harvest season seeded an underground revolution against industrialized agriculture and shrink-wrapped, processed food. CSAs provided a direct connection to the land and the farmers who grew the food.

Numerous women served as the midwives at the birth of this movement and nurtured it from its beginning, drawing inspiration from the works of the Austrian philosopher Rudolph Steiner during the 1920s and similar models in Europe. Robyn Van En was among the most spirited of the bunch. On her Indian Line Farm in Massachusetts, she launched one of the first of two CSA farms in the country, both of which are still thriving today. After her untimely death in 1997 at age 49, her contributions live on in the name of a national CSA clearinghouse of information: the Robyn Van En Center for CSA Resources, housed at Wilson College in Pennsylvania.

Elizabeth Henderson, author of *Sharing the Harvest: A Citizen's Guide to Community Supported Agriculture*, which is considered the "bible of CSA," started the book project with Van En, completing it after Van En's death. Throughout the 1990s, Henderson took the CSA show on the road, speaking on the idea of CSAs. Her passion spoke to many beginning women farmers who embraced the CSA concept and ran with it on their own. The CSA concept resonates strongly with women farmers, stemming from the underlying philosophy of collaboration and community building.

"Lots of women are drawn to and can do outstanding work through CSA," explains Liz Henderson, now retired from farming. "You can live a good life and run an economically viable business through this model."

Additionally, the CSA structure doesn't require heaps of capital or infrastructure; so it serves as an accessible and affordable means for women to get operations up and running. CSA also helps tremendously with cash flow as you receive member payments up front, prior to the harvest, which is when you need the capital for seed and other input investments.

"The majority of our coalition farmers — over 51 percent — are women farmers," shares Erika Jones, executive director of FairShare CSA Coalition, a nonprofit championing CSAs in the Madison, Wisconsin, region. Nationwide, the CSA movement has snowballed to over 12,600, according to the latest numbers available via the 2012 Census of Agriculture. In 1990, that number was estimated to be 60. "FairShare's success really comes from the connections between consumers and farmers working together to grow the movement," adds Jones. This has enabled them to partner with health insurance providers to offer rebates and with local businesses who offer CSA pick-up sites as well as with other community supporters. "This is a win-win both for women and all CSA farmers as our markets increase tremendously."

Women farmer networks

"Feelings of isolation were common as we were all in our way trying to create a different, more sustainable means of farming and feeding our communities, but we were outliers in a male-dominated agricultural system," shares Denise O'Brien, an Iowa organic farmer for over 40 years. Motivated both by the 1980s farm crisis and the lack of female voices at the leadership decision table on national and state levels, in 1997, O'Brien launched the Women, Food and Agriculture Network (WFAN), to serve as leading national connector for women involved in every facet of sustainable agriculture.

How She Sows it:

Denise O'Brien, Rolling Acres Farm, Atlantic, Iowa

Stereotype Smasher:
Four Ways Women Instigate an Agriculture Revolution

Quick trivia question: What's the second verse to *The Farmer in the Dell?* Anybody?

Here you go:

> *The farmer takes a wife,*
> *The farmer takes a wife,*
> *Hi-ho, the derry-o*
> *The farmer takes a wife.*

Talk about stale lyrics in dire need of an update. We should be teaching kids something more like this:

> *The wife took over the farm.*
> *To the land she did no harm,*
> *Hi-ho, times change, you know,*
> *These chicks can really grow.*

Denise O'Brien of
Rolling Acres Farm.

Consider Iowa organic farmer, Denise O'Brien, chief female farmer stereotype smasher. Founder of the Women, Food and Agriculture Network (WFAN), she has been organizing and promoting the voice of women in agriculture since she founded the Rural Women's Leadership Project with PrairieFire Rural Action, Inc. in 1985. "Finally, the tides are starting to turn for women farmers as policies start to slowly change and our economic impact is realized," explains O'Brien. "There's still much we as women, from growers to grocery shoppers, can do to create a healthy food system for future generations."

O'Brien's agriculture and activism roots run deep. "As a young feminist activist and world traveler in my twenties, I didn't think I would return to live in Iowa," O'Brien reminisces. That changed on a trip home when she met Larry Harris, now her husband, who opened her mind to farming in harmony with the earth. The duo launched Rolling Acres in 1976 in southwest Iowa, caring for a cow-calf herd and switching to dairy in the 1980s, then mixed vegetables and poultry for their CSA and various markets.

"When we started farming, we couldn't even say the 'O-word' (for organic) out loud, but we're out now."

Her agriculture advocacy work started during the farm crisis of the 1980s, when an estimated 200,000 to 300,000 family farms were forced into foreclosure. "It was the women on the farms who saw this economic gloom on the horizon, since they often kept the accounting books. They could see where things were heading," O'Brien recalls. "But the problem was that these women still considered themselves 'just farm wives.' Nobody took their voices seriously. We women needed to get out of the role of just making coffee at meetings."

This launched O'Brien on the road to agriculture activism, balancing multiple roles: farming, raising children, and serving as a national advocate and spokeswoman for women in agriculture. "I realized during the farm crisis that if I didn't step up to the plate to organize women, if I didn't instigate something, perhaps no one would." In 1997, she launched the Women, Food and Agriculture Network to help women farmers develop the organizational and speaking skills to collectively advocate for a stronger voice.

"Throughout history, women in agriculture have been relegated to places of assistance versus at the decision making table," O'Brien explains. "It is up to us as women to collaboratively support each other while challenging the system." O'Brien offers some steps to get started:

1. **Meet your female farming neighbors**

   Start coalition building by connecting with others in your community who share your healthy food values. O'Brien first met many local female friends by starting a food co-op. It drew people together with a desire to purchase natural foods not available in their rural area. "Turns out we shared other interests as well, such as politics and organic food," adds O'Brien.

2. **Be Available**

   Make yourself available to attend meetings, particularly those traditionally dominated by men, such as local planning and farm service meetings. When her children were young, O'Brien would often attend meetings with her kids in tow. "I wanted my kids to grow up realizing the importance of being involved and active with the issues you consider important."

3. Push past the expected

"Shake up your own life path occasionally, even if it means not farming a season or two," she adds. "If a situation serendipitously presents itself to you and your heart insists you embrace it, go with your intuition."

O'Brien speaks from experience as she served from 2011 to 2012 as a USDA adviser in Jalalabad, Afghanistan, near the Pakistan border, supporting the region in re-establishing local infrastructure to support small-scale farmers. "This experience both provided me a tangible means to contribute to sustainable agriculture globally and personally challenged me way beyond my current comfort zone," shares O'Brien. "I came home recommitted to the farm, and my next chapter in activism into my sixties and beyond."

4. Run for office

"We need more women in elected positions who hold decision-making power to truly create long-term change." Again, O'Brien puts her money where her mouth is. Well, she may not have had the cash coffers of other candidates, but in 2006, she still ran as the Democratic candidate for Iowa secretary of agriculture. Though she didn't win, O'Brien shook up the political establishment. Despite being outspent two to one on campaign ads and receiving a barrage of negative publicity from agribusiness interests in the final days, she came within two percentage points of winning the election over a conventional corn and soybean farmer.

"The movement is just a toddler," sums up O'Brien. "We've come a long way, but we're not there yet, especially with representation of women farmers on the national leadership platform. But these barriers will only make us fight harder."

Regional networks popped up to geographically connect and support women farmers who shared the value of small-scale and sustainable agriculture. One of the first, and now largest, is the Pennsylvania Women's Agricultural Network (PA-WAgN), launched in 2003 by women farmers, Extension educators, and researchers. Based out of Penn State University, PA-WAgN flourishes thanks to this partnership of higher education resources and women farmers at the grassroots level. PA-WAgN also embraces the fact that women want to run their own farm operations and need the skill sets to do so, right down to tractor maintenance and safe chainsaw operation.

Through extensively surveying women farmer attendees, PA-WAgN crafted the programming distinctly for women and by women. Not surprising, women favored hands-on, practical learning taught by their peers. This research also found that not only did this format substantially increase farm knowledge, the networking and connection aspect proved highly valuable. More than three quarters of the female attendees at PA-WAgN programs surveyed indicated they would stay in contact with someone they met at a workshop as well as take action to create learning opportunities for other women farmers.

In regions without Extension-driven and -funded programming, women in sustainable agriculture started self-organizing, based on an informal assessment of needs and expertise. For example, the Wisconsin Women in Sustainable Agriculture Network (WWSFN) launched in 1994 and ran for nearly a decade, organizing on-farm workshops and training events and connecting with each other via email communications and print newsletters.

"Before today's wealth of online resources and connection opportunities, a network like this was the only option where women farmers could both meet and learn from each other in our state," recalls Jody Padgham, a poultry and sheep farmer in northern Wisconsin who credits the group with empowering her to purchase a 60-acre farm in 2001. "A lot of us were just starting out or new to farms, and had no place to go to learn. Meeting through WWSFN, we shared what we were learning with each other. The safe, women-focused environment was valuable and strengthening."

Always evolving, this original Wisconsin network reinvented itself under my watch as the MOSES Rural Women's Project, an ongoing, multifaceted training and resource program for women farmers, stretching beyond Wisconsin's borders to other Midwest states. With websites, email forums, and Facebook pages, we have greater opportunities to

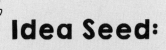

Idea Seed:

"If you're an aspiring or beginning farmer, your first task should be to network, network, network! Go to meetings, join organizations, and meet the folks who have already gone down this path ahead of you. Work with or for them, if you can. Get as much advice and experience as possible before you launch your own business."

— Leigh Adcock, former executive director,
Women, Food and Agriculture Network

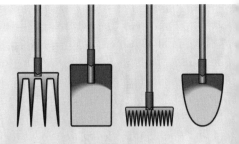

Tool Shed:

Connect and Collaborate through Women Farmer Networks

Find kindred female farmers across the country and on the country road next to you. Below are several women farmer networks committed to small-scale, sustainable, locally focused agricultural enterprises:

Women, Food and Agriculture Network (WFAN)
www.wfan.org
WFAN runs a variety of women farmer and landowner programs, including a national annual conference. There is no membership fee, so be sure to sign up for their e-newsletter and, if you want to access an ongoing national community for women in sustainable agriculture, sign up for their email forum where you can post questions and resources.

Women's Agricultural Network (WAgN)
Several Extension-run WAgN programs run primarily in east coast states, often including teams of regional farmer representatives organizing localized events.

Pennsylvania WAgN
agsci.psu.edu/wagn

Vermont WAgN
uvm.edu/wagn

MOSES Rural Women's Project
mosesorganic.org/projects/rural-womens-project/
MOSES offers various on-farm, women-farmer-led workshops and programming, including local networks in Wisconsin.

keep connected virtually, yet we still prioritize face-to-face connection opportunities whenever possible.

Reclaiming the farm and new domesticity

While *Soil Sisters* focuses on farm-based, entrepreneurial livelihoods, this back-to-the-farm movement parallels a renewed interest in domesticity. Whether you're a Shannon Hayes version of a radical homemaker or earning a knitting merit badge through farmgirl sisterhood at MaryJanesFarm, more women are shelving the Michael Kors handbag and making their own hand-felted version — with fiber from their own sheep! We've embraced "homemade"

Women farmers networking at "In Her Boots" workshop with farmer Audrey Arner of Moonstone Farm, Montevideo, Minnesota.
JOHN D. IVANKO

with vigor, and yearn to do everything from scratch, from kid-raising to kombucha-brewing.

Shannon Hayes runs a grass-fed beef operation, Sap Bush Hollow Farm, with her family in upstate New York. Like many in the back-to-the-farm movement, she rejects mainstream consumer culture to live the lean, green home-based DIY lifestyle. Her initial intuition that she wasn't alone as a "tomato canning feminist" proved true when media started calling. "On the Farm, a Rat Race in Sheep's Clothing," read the headline by the *New York Times* that brought the discussion of renewed domesticity into the mainstream of Manhattan. Her mailbox and email folder overflowed with stories of others in their quest for homespun independence. Several road trips later, these stories evolved into the fuel for *Radical Homemakers*, which concludes that social change begins at home.

"Reclaiming our domestic skills is the starting point; our continued happiness, creative fulfillment, and further healing of our society and planet requires that we look beyond the back door and push ourselves to achieve more," explains Hayes. "It is not enough to just go home and put down roots; we must also cultivate tendrils that reach out and bring society along with us. Women launching farm businesses are a natural extension in that these operations do more than just bring in cash income, they can serve as teaching and educational tools to change the food system in your community and, eventually, the world."

In her book, *Homeward Bound: Why Women Are Embracing the New Domesticity*, author Emily Matchar defines this "new domesticity" as something that speaks to "deep cultural longings and [a] profound shift in

Lisa Kivirist cooking on the woodstove at Inn Serendipity.

JOHN D. IVANKO

the way Americans view life." It involves activities, particularly embraced by those in their twenties and thirties, that address our "longing for a more authentic, meaningful life in an economically and environmentally uncertain world." She says earlier, "Our new collective escapist fantasy is more likely to involve a Vermont farmhouse and a cute Anthropologie apron than a SoHo loft and pair of Manolos."

This DIY domestic front proves to be fabulous frugal inspiration for crafting your farm-based lifestyle. No need to buy something when you can learn to make it yourself.

During the winter months, my family plays this "eat through the pantry and freezer" game; we stop buying food from January through late spring, when fresh items start popping up in the fields. This challenge motivates us to use up the vegetables we froze and preserved the previous summer. We cook creatively through the winter, turning out spanakopita with our spinach or moussaka with frozen tomatoes and eggplant, plus potatoes from the root cellar. Many of my recipes in *Farmstead Chef* came directly from experimenting in the kitchen. This cook-at-home commitment showcases the frugality and freedom aspects of this new domesticity.

Likewise, reusing and recycling items versus buying new creates economic independence. During the B&B start-up phase, I read various "How to start a B&B" books where the authors suggested budgeting at least $10,000 per room for start-up costs. The only new items we purchased were mattresses for the two guest bedrooms. Everything else in the farmhouse came free, most often hand-me-downs from friends and families with overstuffed closets or spare rooms. Self-taught at refinishing and painting, my family and I spruced up the farmhouse. Wabi-sabi, shabby chic you might call it. But domestic DIY let us launch the B&B debt free.

Back to the future: women farming today

Cue the new rock stars of agriculture: women! With more women continuing to enter agriculture as new farmers, the media have acclaimed women

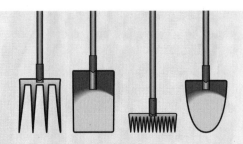

Tool Shed:

Domestic Farm Divas Unite!

The renewed domesticity movement comes in varying perspectives and priorities, with a handmade bucketful of recipes and resources to read over your glass of home-brewed kombucha:

Books:

Radical Homemakers: Reclaiming Domesticity from a Consumer Culture and *Homespun Mom Comes Unraveled ... and Other Adventures from the Radical Homemaking Frontier,* by Shannon Hayes

MaryJane's Ideabook, Cookbook, Lifebook for the Farmgirl in All of Us, by MaryJane Butters

Homeward Bound: Why Women Are Embracing the New Domesticity, by Emily Matchar

The Hip Girl's Guide to Homemaking: Decorating, Dining, and the Gratifying Pleasure of Self-Sufficiency on a Budget! by Kate Payne

celebrities in the field with such headlines as "Breaking the 'grass ceiling': More women are farming" in *USA Today* and "Old McDonald Might Be A Lady: More Women Take Up Farming" on National Public Radio.

USA Today and NPR are onto something statistically. According to the US Census of Agriculture, there are nearly one million women farmers today, representing about 30 percent of all farmers. About half of us call ourselves the boss; we're the "principal farm operator" by USDA terms. This means we're the person in charge of the farm's day-to-day operations. Together, we add up over $12 billion in annual sales of farm products. Our numbers have nearly doubled over the past 25 years, which makes us the shiny, happy bright spot amid rather dismal statistics that often reflect the continual decline of the total number of farms.

New women farmers are primarily starting small-scale, diversified, locally focused and family-run operations, bucking the corporate agriculture trend of fewer small family farms and more consolidation. According to the Census of Agriculture, there are 750,000 fewer family farms than in 1981, adding up to one million family jobs lost. While the number of family farms decreases, the average acreage of farms increases as farms consolidate under larger corporate entities. Over the past 70 years, seven million farms have become two million.

How are we, as women farmers, changing the face of American agriculture? Per the Census of Agriculture, our farms are definitely smaller than the average: 54 percent of our farms are smaller than 50 acres in size. For farms with a female principal operator, that number jumps to 82 percent. We gross lower total sales as well. Over 75 percent of farms grossed less than $50,000 in 2012. For women-owned farms, that number jumps to 92 percent.

But don't think we're just getting by. A struggling blip on the farm business scene we're not. We're more likely than our male counterparts to raise a diversity of crops, including vegetables, fruits, and increasingly, flowers, herbs, and other specialty crops. We're also more likely to own our land and sell our harvest directly to our customers. In many ways, we're carrying the agricultural torch of our Depression era foremothers, except in this round the dollars earned and businesses created are all ours. Most of what we grow on our farm we can eat — usually fresh.

Contrast this with our farming neighbors who, with plentiful USDA subsidies, grow commodity crops, most of which are turned into animal feed or fuel. Yep, acres of corn, soybeans, and alfalfa cover rural countryside throughout the nation — crops rarely consumed as actual human food. In fact, 40 percent of corn production is turned into ethanol that Americans burn in their cars, SUVs, and minivans.

As a group, we champion land stewardship through every compost bin we start and tree we plant. We reflect this conservation value in how we speak about our farms. WFAN studies have found that women communicate differently about farmland. They found — no surprise — that we refer to our land as a community resource rather than a commodity crop. Our farms are not widgets that can just be bought and sold; rather we see our land as something bigger than ourselves, something that we want to share with those around us.

But the data from the Census of Agriculture can quickly get murky and complicated. It tracks more than just a total count of warm farmer bodies; its data includes statistics on operations ranging from small family farms to large farms; information on young, beginning farmers and older, experienced farmers; and insight into traditional, rural farming versus trends in areas such as lifestyle and urban farming.

When it comes to women in ag, there's misrepresentation of principal operators, who for various reasons don't participate in the survey. To the credit of NASS, changes are in place to not only better account for women but to have more accurate numbers on new farmers. In addition to women, groups including Hispanic and Hmong farmers are also on the rise. The Census started officially counting women in 1978, but it wasn't until 2002 that it included space for more than one name as principal farm operator. For the vast majority of family farms, the male head of household's name went in the one spot before the change. Even if the woman played an important and equal farm role, she did not count, according to the census takers.

In an effort to better represent these women and other minority farmers, the Census of Agriculture now includes a place for respondents

Idea Seed:

"As women farmers in our male-dominated culture, especially in agriculture, where bigger is deemed better, we need to be seen and respected as contributors to our local economy. As women we earn respect through our skills and success as business owners, but we need to step to the leadership plate together to make our voices collaboratively heard."

— Pakou Hang, founder of the Hmong American Farmers Association, a nonprofit organization providing a voice for a minority farming community and enabling Hmong farmers to best blend their cultural farming practices to produce products for the American market

to list the total number of operators on a farm. The "farm operator" is defined as "the person who runs the farm, making the day-to-day management decisions." This person doesn't necessarily have to be the actual farm owner. Additionally, there is now space to list more detail on up to three operators. This additional information includes gender identification, which is the place to include female partners such as wife or siblings, or farms that may be operated by two women.

Further complicating the Census, many new women farm owners may be unaware of the Census, feel they run too small an operation to be counted, or become overwhelmed in completing the form and give up. For Census purposes, a farm is "any place from which $1,000 or more of agricultural products were produced or sold, or normally would have been sold, during the Census year." It's still a "one size fits all" survey and most definitely geared toward large, conventional operations. It can be a challenge to wade through over 20 pages of questions, trying to make your half acre of herbs or pastured poultry operation fit the check boxes. Agritourism ventures, such as farm stays or tours, are not listed as a specific farm businesses category on the Census. While there is a category titled "other" income, which results in farms entering a wide range of sources, it dilutes the validity of entities such as farm stays like ours that generate viable agritourism income of $10,000 to $20,000 a year.

For better or worse, data from the Census of Agriculture drive decisions, particularly how funding and resources are allocated in the Farm Bill, the comprehensive omnibus bill passed about every five years that is the primary agricultural and food policy tool of the federal government.

Legislators use this data when shaping farm policies and programs, which is why we need to fill out that paperwork and make our female-operated farms count. More accurate numbers increase our influence and priority status in both federal and other funding opportunities such as private foundations donating to nonprofit groups.

Reality check: history of discrimination

The inspiring growth of women farmers comes after decades of discrimination against women farmers in our nation's agricultural programs and in access to resources. Women farmers have experienced inequities based on gender when applying for federal farm loans or other agriculture programs. Sometimes such discrimination was as direct as being told to come back with spouse or father.

This prompted a class action lawsuit, *Love v. Vilsack*, filed in 2000 by a number of women farmers against the USDA for gender discrimination in the administration of the USDA's farm loan programs. This resulted in a settlement claims process through which women farmers who had experienced such discrimination could obtain a financial payout.

Fortunately, the USDA is currently taking strong action to both acknowledge past issues and rectify misdoings for both women farmers and other underserved populations, such as Native Americans and African Americans. In 2011 the USDA opened a claims process for women farmers to apply for compensation if they felt they had been improperly denied farm loans or other benefits because they are female. A similar claims process is happening on behalf of Hispanic farmers.

For some USDA programs under the Farm Bill, women qualify as a "socially disadvantaged (SDA)" farmer group, similar to other minority farmer groups such as African American or Native American. The SDA status brings priority funding in some programs, such as beginning farmer loans or training grants to nonprofits, through the Beginning Farmer and Rancher Development Program (BFRDP). We'll look more into these resources in Chapter 4.

"We still hear stories about women going into USDA offices and being asked 'where is your husband?'" shares Bridget Holcomb, executive director of WFAN. "It's great that the USDA is prioritizing women and other socially disadvantaged groups in agriculture, but the real change will happen when we have more women farmers and more women in elected offices."

Mantras of meaning

Blending this past with our farming future, you'll find several key themes throughout this book.

Land stewardship and sustainability

Most of the women you'll meet in these pages champion caring for the Earth by employing sustainable and organic farming practices. No surprise here. We're not in farming to make a fast buck or grow big and industrialized. We plant cover crops, make compost, and smile when we find earthworms when we dig up our potatoes. We farm with the future in mind, eyeing the world with the "seventh generation" perspective. How is what we're doing affecting the planet seven generations out? We plant oak trees for our grandchildren's children to climb. We never use poisons on our land.

Education

That's not just a tomato — it's a teaching tool for transforming our food system. Our mission goes beyond a financial transaction. We don't just trade cash for cucumbers at the farmers' market. We want our customers to identify and embrace the choices they have to shop with their values. Every time we slice a tomato or cucumber for someone to taste, we're optimistic that their taste buds will tingle and cranial light bulbs go off that connect how local and farm direct make the flavor difference.

Partnerships, community — and men!

Though we celebrate and champion female farmers, few of us farm alone. Some may have partners or husbands with whom we plow till the end of the row. Others are the sole owners of their operations but have hired hands, seasonal help, interns, CSA members as work sharers, or a combination of these. Isolation can be highly unappealing because we're naturally drawn to community and collaboration. Appreciating help from everyone — like neighbors who bring over their horse manure and adult siblings who dedicate their vacation time to come help with the harvest — *Soil Sisters* gratefully celebrates the idea that none of us has to do this alone. A special shout-out to the men in our lives, whether husbands or brothers or that mechanic in town who never made us feel inferior, even on that first day we walked into his shop.

Creativity: farming as art

Think of female farmers as artists. Our farms are not just productive fields

Idea Seed:

"Even weeds, if that name suits you, like dandelions teach us something. Just look at them, arriving in your garden soil with their fat roots, aerating it and giving minerals to feed the dirt. Then they shoot up a glorious yellow flower that makes the best addition to fresh salads and biscuits (yes, the flower is edible). Then when they go to seed their perfectly round wish balls are harvested with reverence by children, even the ones housed in our adult bodies, and are told the dreams we have for the world that has yet to come. After all that good medicine that the dandelion displays to us, we are taught to eradicate it. So we might pull them out one by one, or pave them over with cement, or develop toxic chemicals to burn them up on site, cursing them all the while. But do you think that gets rid of them? No way! Because when you are carrying big medicine for the world, then no matter what anybody tries to do to get rid of you, you don't let them! That is the gift of just one, often uninvited plant. Teaching us perseverance and persistence."

— Valerie Segrest, member of the Muckleshoot tribe and project coordinator of the Muckleshoot Food Sovereignty Project

but landscapes of beauty. An inherent artistry exists in what we do. We see and foster beauty in our surroundings and landscape. For example, you'll meet Gabriele Marewski of Paradise Farms in Homestead, Florida, who paints every door on her farm pink simply because she likes how it looks — and she can.

Likewise, this concept of art applies to our farms serving as our canvases, blank slates on which we can create and cultivate our livelihoods. You'll find diversification a common theme among the women farmers in this book. While this makes practical sense from a business management perspective — having multiple income streams helps mitigate risk — diversification also provides us a creative outlet. As women, we're natural multi-taskers and thrive when one idea can feed the next. Having the opportunity to nurture different ideas from our farm-home base can empower us and build confidence to keep thinking big and connecting the dots.

The pink doors of Paradise Farms, Homestead, Florida.
JOHN D. IVANKO

Collaboration

A 2010 University of Wisconsin Extension research study supports this idea that women farmers learn best from each other. Women farmers, in both traditional dairy and small-scale diversified crops, were asked where we go for agricultural information. After our fellow farmers, our second information source is grassroots nonprofit organizations like MOSES. More traditional sources of information, such as the Farm Service Agency (FSA) or a county's Extension agricultural agent (most of whom, not surprisingly, are male), fall low on the trusted information totem pole. This lack of connection to agencies may also be due, in part, to the fact that many of us new to agriculture simply aren't aware of these readily available resources.

"Hands down, women turn to other farmers as their primary source of trusted information," explains Dr. Astrid Newenhouse, senior scientist at University of Wisconsin-Madison and a lead researcher in this study. "The results of our study are clear that female farmers learn best from each other and they champion opportunities for women to connect and share resources and experiences." Studies in other states and regions have shown the same thing.

There's no one hundred percent right way to do anything related to farming. This book's approach is to pepper you with various inspiring perspectives and experiences. Use these as a foundation and launch pad to ask more questions and garner insight, especially from women farmers in your neck of the woods familiar with your growing climate, market, and local scene.

Idea Seed:

A Woman's Hands

Her hands are cracked, reeling from the whipping wind. Split, torn and dotted with splinters, numb to the impact of the wooden shovel. Cuts fade and reappear, garnishing her knuckles. Her palms tell the story of the day's work, etching out the lines with black soil to the edge of her fingers, retracing every inch of land ploughed, every seed planted. Her forearms are brushed with dried mud, some splattered onto her face. The rest is caked in every crevice of her fingernails and painted onto her faded, tattered clothes. It's too early to see the calluses on her palms, but if you were to hold her hand, you'd feel them.

— Natasha Bowens, author of *The Color of Food*

Chapter 2

You as a Farmer

MULLING OVER LAUNCHING A FARM ENTERPRISE? Chances are it's not because someone planted that idea in your head and said, "You'd be a great farmer." The idea didn't start flashing during some career assessment screening or because a college counselor proposed the idea. It likely didn't come up around the Thanksgiving dinner table. Becoming a farmer probably sprouted inside your heart and soul. You smiled every time you thought about chickens, compost, or cabbage plants.

We female farmers dance — and dig — to our own drumbeat. We're mavericks in muck boots, wanting to create our life and farm by our own rules. Whether we're reconnecting with our family's agrarian roots or plowing new ground, like I am, we share a sisterhood with Mother Nature and wildly want to care for and play in her soil, plant her seeds, harvest her abundance, and transform her sustenance into healthy, nourishing meals for our families and communities.

Given that few of us were guided to farming, many of us probably followed detours to places with paved streets, sidewalks, and Starbucks. We landed jobs in cubicles or in business bureaucracy that sucked us dry, leaving us feeling like a round peg in a square hole. A clear escape route and destination were clouded by parents' aspirations, social stigmas associated with being a "dirt farmer," or our own discomfiture at not working in the field dictated by college degrees.

In this chapter, we'll explore what it means to be a farmer, and look at the characteristics needed to succeed. By better understanding ourselves, strutting our strengths, and overcoming our weaknesses, we grow into stalwart farmers.

Gardener to farmer: hobby versus business

The majority of aspiring female farmers already have home gardens. From coaxing along a tomato plant in a container on a deck to cultivating abundant backyard plots from which you're keeping the whole neighborhood supplied with zucchini, the time you spend weeding the carrots or harvesting the first Early Girl tomatoes brings joy beyond anything else you do. When you give a bag of garden bounty to your neighbor and see that smile, life is divine. For many of the women I've interviewed and worked with, it answers a spiritual quest: What on God's earth am I to do with my life? Grow the food that feeds my family and community.

But a farm business isn't a gardening hobby on steroids. The Internal Revenue Service (IRS) outlines specific guidelines in determining whether something is a hobby versus a business:

- Does the time and effort spent indicate an intention to make a profit?
- Does the taxpayer depend on income from the activity?
- If there are losses, are they due to circumstances beyond the taxpayer's control or did they occur in the start-up phase of the business?
- Has the taxpayer changed methods of operation to improve profitability?
- Does the taxpayer or his/her advisers have the knowledge needed to carry on the activity as a successful business?
- Has the taxpayer made a profit in similar activities in the past?
- Does the activity make a profit in some years?
- Can the taxpayer expect to make a profit in the future from the appreciation of assets used in the activity?

While romantic dreams of growing produce as part of your livelihood may be alluring, crossing the street to the business side requires shifting to a business mindset. Those neighbors will need to start giving you greens, as in dollar bills, for your collards and cucumbers. That means you'll need to shift your perspective and embrace it. Your first customers will be those people already connected to you.

There are practical, strategic reasons to be a legitimate business: deductible expenses. By operating as a business, you now track your supplies and tools as expenses under IRS guidelines. Your business doesn't need to generate all your income to be legal in the eyes of the IRS. You can start as a side business selling at one farmers' market to test the waters while still holding down your day job, for example. That's a savvy strategy for start-ups. The bottom line is that even a small venture needs to eventually turn a profit. Per the IRS, the business must earn profits at least three out of every five years.

Note that you can't deduct personal, living, and family expenses as part of your business. The IRS defines personal as "rent and insurance premiums paid on property used as your home, life insurance premiums on yourself or your family, the cost of maintaining cars, trucks, or horses for personal use, allowances to minor children, attorneys' fees and legal expenses incurred in personal matters, and household expenses. Likewise, the cost of purchasing or raising produce or livestock consumed by your family is not deductible." We'll cover more on business expenses in Chapter 7, but for now, make the mental shift to start thinking like a small business owner, and turn your love of the land and growing food into a viable livelihood.

Up until now we've been looking at things from a small-business perspective, like the diner or mechanic shop in town. While business fundamentals are important and form the bread and butter core of your operation, embrace your inner entrepreneur. The business owner manages the fundamentals, but the entrepreneurial mindset asks, What if? It's the entrepreneur that leads social change and asks the big questions such as, How can we make organic affordable for all? What could I do that gets kids excited about eating these vegetables? Can a group of my local female farmer friends and I buy a food truck to host farm-to-table dinners and structure it as a cooperative model that other farmers could replicate? Think like an entrepreneur and both your business and our world will benefit.

Soil test: skills assessment

Good news: Regardless of your background, you start with some transferable skill sets. Whatever your life experiences or career path, which may feel like light years away from farming, tangible skills transfer to your agricultural enterprise. Plant pride in the things you already know. Confidently train yourself on what you don't.

Idea Seed:

"Set your goals early in your planning. Always be honest with yourself — if something isn't working, change it. Celebrate every milestone, and never be afraid to laugh at your mistakes — there will be plenty of them."

— Mary Peabody, Women's Ag Network-Vermont

Tool Shed:

Understanding *Publication 225, Farmer's Tax Guide*

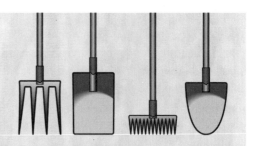

This publication explains how the federal tax laws apply to farming. You are in the business of farming if you cultivate, operate, or manage a farm for profit, either as owner or tenant. A farm includes livestock, dairy, poultry, fish, fruit, and truck farms. It also includes plantations, ranches, ranges, and orchards.

www.irs.gov/Publications/p225

For those in business, nonprofit, or governmental sectors, numerous skills will serve you well in farming:

Business or organizational planning and administration

Whether you're a brand manager for some widget or an administrative aide in a state agency, you probably have been exposed to various aspects of running a business or organization. Sure, manufacturing a widget requires a completely different process from cultivating cucumbers, but you have skill and experience in the thought process and planning that will be the bones of your successful farm venture. Whether developing production timelines or calculating cost inputs, managing staff or processing customer orders, in many ways, you've been training for years to own your own farm.

Marketing

Any time you sold or convinced someone of anything, that's a skill your farm venture needs. If you're a waitress who can vividly describe the daily specials to snare an order, that's exactly what you'll be doing to lure someone to your market booth to sample your heirloom tomatoes. Understanding branding and crafting a message behind your product enables you to readily identify and sell your farm story. You're not selling honey, you're offering potential customers the ability to "bond with the bees" and "know your farmer, know your food."

Writing and communicating

Effectively communicating your wares through the written or spoken word is probably something you've been doing for years. The ability to synthesize information into an annual report transfers to writing a farm business plan or crafting a grant application. Likewise, oral skills play an important role. For example, if you diversify into agritourism and give farm tours or teach workshops, you're drawing on the skills you used if you worked in a middle school classroom.

People skills

Have a knack for navigating and resolving office drama? That ability shifts to providing feedback to staff and keeping everyone motivated and positive on those August harvest days when the humidity index hits sweltering. Those people skills come into play on a personal level, too, when learning to work with your spouse or partner. From divvying up the chore lists to strategizing the big picture for the farm, it all boils down to relationships.

Social capital

Whatever network and connections you currently have, some can transfer and support your farm. Perhaps you know people skilled in certain areas where you need expertise, such as designing a logo or creating a website. Take these friends out for coffee, share your dreams, and ask for their advice. Depending on geography, your former office site could be your first CSA drop-off point. People who know and support you will be your biggest champions.

Financial capital

Whether in savings accounts, stocks, or a 401(k), the dollars needed to fuel your farm venture may be at your call to get your farm off the ground. If you've been prudent in your savings, that nest egg may launch you to your next life chapter.

Farm transplants: the right fit

This new movement of women farmers reflects the core pillar of sustainable

Idea Seed:

"Farming will leave its mark on you. You'll be physically stronger, more emotionally flexible, better able to solve problems and more resilient. Thinking on your feet every day will give you the confidence to handle whatever life throws at you."

— Catherine Friend, author of *Hit By a Farm* and *Sheepish*, Rising Moon Farm, Zumbrota, Minnesota

agriculture: diversity. Our farm businesses represent a dizzying array of products and services, which is why we struggle in completing the Census of Agriculture. We don't fit into a few simple checkboxes. We don't raise corn and soybeans. We raise multiple varieties of fruits, vegetables, and herbs. We may have goats, chickens, ducks, sheep, and a herd of beef cattle. On a personal level, we come from a variety of backgrounds, interests, ages, experiences, and cultural groups. Some of us grew up on farms or come with strong connections to family land. Others, like me, grew up with no direct ties to agriculture and discovered our passion for producing, preparing, and sharing food later in life.

After years of interviews, workshop field days, and women farmer trainings, I've distilled women farmers into four distinct categories. Which best describes you? Or perhaps you are a combination. This book shares stories you can relate to. There's probably a woman in your community to connect with as well.

1. Encore farmer: agriculture as a second career

Encore farmers enter agriculture after one or more careers in other fields and may be mid-life or even older, nearing retirement. If you're in this category, very likely you have no direct farming experience. Perhaps it's a dream you've been harboring. Maybe you want to raise your family far from the rat race. Or perhaps your kids have fledged and you can finally focus on what you want to do.

As an encore farmer, you bring seasoned life experience, no matter what the industry.

You possess a reality check you probably didn't have in your twenties or thirties. You know that a Plan B, plus some cash reserves, is wise. Speaking of reserves, you may have amassed significant savings and investments over the years or an IRA that you can borrow from. Be sure to check on how to strategically tap into your IRA account without any early withdrawal penalties. On the topic of penalties, if you receive social security payments, research and understand any income thresholds that will result in increased tax liability. Run your farm in such a way that any income goes back to your business, not unnecessarily flowing to the coffers of Uncle Sam.

Time is of the essence to you. No longer the youngest chick in the flock, you come to farming with a heightened sense of urgency. This could be your one shot at creating a farming livelihood and land legacy. Life quickly marches forward, and you want to get this farm up and running while you still have the physical ability to do it.

Embrace that physical reality: a forty-, fifty- or sixty-something body is not the same package as you had in your twenties — especially if the majority of your waking hours the past few decades have been spent parked in a cubicle. The physical realities of farming will need attention and training so that you achieve your personal best health. This may also mean understanding and accepting your physical limitations. You can't do it all. Hiring support staff and organizing your business plan around your physical abilities will go a long way in fostering success.

Time, or the lack thereof, can be a motivating concern as you plan your farm business. Some crops, like apple trees, take three to five years to mature; do you have that kind of time to invest? Or can you find a crop with market appeal and a shorter maturation cycle, such as a strawberries?

If you're worried about your age, you'll be pleasantly surprised at the diversity of ages represented in this new farming movement. You'll fit right in. The White House's 2013 *Economic Report of the President* calculates that "the average age of U.S. farmers and ranchers has been increasing over time." One third of beginning farmers — defined by the federal government as having been in business fewer than 10 years — "are over age 55, indicating that many farmers move into agriculture only after retiring from a different career."

"While I was on the older side of the age range of attendees, we shared this common passion for launching a farm and shared similar goals and needs," shares Pam Walgren about her experiences attending the New Farmer Summit, organized by MOSES and Renewing the Countryside. Walgren runs Perennial Journey Farm in southwest Wisconsin, growing vegetables and selling at the local farmers' market. "There actually was a lot of good cross-pollination and collaboration between beginning farmers 40 years apart in age, a spirit that we're on the same road together and support each other."

Sooner than later, you'll need to take that entrepreneurial leap of faith and press the start button. This process can be daunting and a bit ironic: You so want to leave your current unfulfilling job and enter the world of sheep in the barn and stars in the sky, but the security of that weekly paycheck seems to tether you to the earn-and-spend W-2 job culture. There's an appealing comfort and addictive security in routine. It's easy to play mind games, such as spending too much time researching and writing the perfect business plan, waiting for the perfect formula that guarantees the success of your venture.

But you know, deep in your heart, that the vision and perfect plan will never all of a sudden fall perfectly into place. Admit it: You're using the research

How She Sows it:

Paula Foreman, Encore Farm, St. Paul, Minnesota

Building a Farm Business Mid-life, Bean by Bean

Paula Foreman with her niece, Margaret, at Encore Farm.
<small>TERESA FOREMAN</small>

Most people plan a big birthday bash to ring in a milestone birthday like 50. For Paula Foreman, that landmark date inspired something way deeper than frosting and sprinkles: Foreman started her farm, entering a new agriculture chapter as she entered mid-life.

After careers as diverse as bread baker, legal advocate, daycare mom, and construction carpenter, Foreman launched her mid-life "second act" as a farmer in 2007 with Encore Farm, a name that can serve as a rallying mantra for mid-life women launching farming careers.

"I named this venture Encore Farm as it's a tribute that fresh starts and new beginnings can happen at any age," explains Foreman.

Foreman exemplifies the female encore farmer in that, empowered by maturity and moxie, she runs her business by her own rules. She doesn't fit a stereotype of what a farmer should be, although she's always had agricultural inclinations. She had tractor toys as a kid and spent weekends on a friend's farm as a teen, jumping in with fencing, haying, and milking. In her forties, she connected with the writings of Wendell Berry. "When I read Wendell Berry, it confirmed farming should be my next act," shares Foreman.

"I didn't grow up in a farm family, don't own acreage, and never even thought of myself as a gardener. Nonetheless, I have always been drawn to a self-sufficient and handmade life," explains Foreman. "Farming ties it all together for me; I've found my calling."

Her first step was harvesting farming experiences for a few years before launching. She held a worker share with her CSA and completed the Land Stewardship Project's "Farm Beginnings" course, a year-long, in-depth program through which you develop your farm business plan.

She knew her business would need to be small and lean to succeed. Though she'd have some support from her husband, Jim, Encore Farm was hers to run with. Foreman chose to specialize, producing one thing very well: dried beans on one acre of leased land, using organic and sustainable practices that include composting and cover crops. Foreman leases an additional two acres that she keeps in cover crops for soil fertility, pest management, and pollinator habitat.

"I grow unique heirloom varietals of beans," Foreman explains. "When I first started seriously thinking about farming, my *Seed Savers* catalog came in the mail with this beautiful photo of dried beans on the cover, and I was immediately drawn to it. This is what I needed to grow." Her beans have a wide range of histories, tastes, and textures, such as Lina Cisco's Bird Egg Beans, which are rich and creamy in taste and approved for inclusion in the United States Slow Food Ark of Taste, a program protecting foods nearing extinction. She sells primarily to chefs in the St. Paul and Minneapolis area, who feature the beans on their fall and winter menus.

Those little dried beans quickly add up to strategic marketing sense, as they are a shelf-stable crop. This makes the business easier to manage than a perishable crop such as lettuce greens, which have to be sold shortly after harvest. She can sell beans year round or until she sells out.

"It's possible to do this on a shoestring budget," sums up Foreman. "Let go of your old ideas of what 'wealth' is all about, and define it on your own terms. Our real wealth is in each other, our ability to grow food, and in living in the life of our choosing. It may look humble from the outside, but from my side of the fence it's rich and satisfying."

phase as a crutch, an excuse, to not take that leap of faith that will trigger a whole series of events that just might be beyond your wildest dreams. Change can be intimidating, even positive change. Ultimately, you'll need to evaluate whether your vision is to create a viable, income-generating farm business or to have a hobby homestead with a few chickens and garden to play around in.

2. Lone ranger: solo female farmers

Lone rangers run solo farm operations sans partners or others directly involved with and responsible for the business. This could be someone married or in a relationship but with a partner in a totally different career, or someone running the farm completely independently. An estimated 2.1 million women run such solo ventures, but they are often overlooked in farm training programs, even training specifically designed for women.

"Life is very different if you run a farm with someone else, whether a business partner or a spouse. I can't relate to people, even women, who tell their stories from the perspective of sharing the workload with someone else. I realize and embrace the fact that I'm on my own, but let's talk about topics from my viewpoint and lens. I need ideas that work for me." This type of comment comes up often in the "In Her Boots" trainings I run for MOSES. I now make sure there's a solo woman farmer represented as speaker or farm host for my workshops.

As queen of the coop and other farmstead operations, you can create and customize everything based on your objectives, goals, and whims. If you're petite, lower the shelves. If you crave the color purple, paint the barn that color versus the expected brown or red. Your farm is one hundred percent an expression of you; take advantage of that fact and plow your own path, literally. Some decisions may be purely practical, like storage based on your height. Some decisions may be aesthetic, based on what you define as beauty. Whatever that may be, relish the fact that you're the queen bee of your hive.

Like a ninja, solo operations can quickly adapt as needed. With no committees or partners to consult, you can make fast decisions that best suit your needs, priorities, and opportunities. This can be a major advantage when you are starting out.

However, the whole term "solo" can be misleading since we're part of a larger web of the sustainable agriculture community, a piece of a broader network. If you're a sole proprietor of your farm business, your "community" will be built in different ways and may be something that you need to seek out and create. From her mechanic to extra seasonal staff support, from loyal CSA members to that non-farming friend she can always call and vent to, a lone ranger chooses mentors and connections. Be strategic in how you bring this network in. Carefully consider who will play a role.

Barter can be an equitable option if you need a professional service and are low on cash and high on eggs, pickles, or whatever farm bounty you may have. "I absolutely love bartering with farmers," shares Kristi Waits of Second Cup Media, a web design and social media consulting company in Vermont. She's traded web design work for meat, produce, and eggs. "It's a double win for me personally when I can barter with a woman farmer because not only do I receive some fabulous farm fare, I'm also able to use my skills to help her business succeed. The process is much more personal and rewarding than the usual cash exchange, and I've become close friends with several of my farmer barter pals."

Understand and embrace your social needs when you're operating solo. Though some of us may thrive in a more introverted and isolated state, others may still need to connect regularly with others. This could mean joining a local civic or business group that meets regularly like the Rotary Club or finding a purely social gathering peppered with people outside the farming scene for some conversation variety like a library book club.

Possibly you have a spouse or partner who's a big part of your life but not directly connected to the farm scene. A partner working in a different career can bring a dose of balance and perspective by having someone at the dinner table not as obsessed as you with rainfall and squashing Colorado potato beetles. Or perhaps your partner with a day job brings in a stable income or insurance or other benefits that empower you to keep on cultivating. This can be especially vital during your start-up phase.

3. Fledgling: young farmer

Ahh, the joys of youth! You have it all: ambition, time, physical stamina, and the energy level of a puppy. You are a sponge for absorbing and learning anything you can about farming, and your résumé of agricultural experience runs wide and deep via apprenticeships, conferences, workshops, self-study. Perhaps you even came out of college with a degree in sustainable and organic agriculture. Most vivid is your vibrant and inspiring mission behind your green acres vision: You're not just looking to earn a paycheck or feed a few families; you want to transform our food system, and you see your operation as the golden ticket.

Katie College, Stoney Creek Valley Farm, Dauphin, Pennsylvania

Sole Operator:
Three Ways to Make It Work with a Non-farming Partner

"Envision a triangle with three points: you, your partner, and your farm. This is the *ménage à trois* of the solo farmer in a relationship," shares Katie College of Stoney Creek Valley Farm in Pennsylvania. A farmer with a philosopher's soul, College built her successful diversified vegetable operation as an encore farmer, her first career being a high school band director and professional French horn player. The philosopher side comes out when asked how she built her farm business as a solo woman farmer while her husband of over 20 years, Greg, held an independent career as an actuary. Both the farm and marriage came out stronger as a result of the triangle.

College describes the three lines of the triangle as one from you to your partner, one from you to the farm, and the third from your partner to the farm. "First and most important, there is one line you can't touch or influence, and that's the line between your partner and the farm. You cannot control your partner's love or enthusiasm for the farm or the farming process. It's out of your hands, out of your circle of influence, if you don't mind another geometrical reference," College adds with a smile. "When you have a partner who doesn't share your passion for farming, there's a tendency to want to manipulate that third line and create the outcome we desire. That's not only impossible, it's unethical, as you're exerting your wishes on someone else."

College's mantra: Your partner is under no obligation to become the person that you want him or her to be. This advice isn't just related to farming, she quickly adds. "It's the whole relationship thing: Keep in mind, your partner is not obligated to change to meet your expectations." What we can do is strengthen the lines connecting us to the two points of the triangle we can touch directly: our partner and our farm. We might have

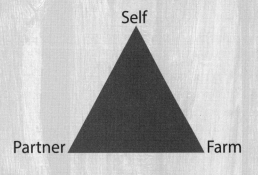

Self

Partner   Farm

a shaky relationship to begin with, or we might, frankly, not be good farmers. "In those cases, we often want to put the blame where it doesn't belong, on that third, untouchable line."

College offers this illustration: We pretend that the problem with our bottom line financially on the farm is the lack of support from our partner, when in fact it might be that we ourselves aren't managing the farm well. We blame the difficulty in our relationship with our partner's disinterest in the details of farming, but maybe the problem stems from a lack of solid communication. "The problems are so often ones that are in our circle of influence, but it's easier to shift blame, in our minds, to the third line, the one we can't control," sums up College. To help manage that triangle and make a marriage or partnership work when only you are the farmer, College offers three tips:

1. Gain perspective by imagining a scenario in which you have a different job entirely. Let's say you are a surgeon. What exactly would we want from your partner in this case? You'd want validation, sympathy, encouragement, and sometimes suggestions about solutions to problems. You're not asking them to jump in and assist.

2. Embrace and own the idea that you're the farmer. Your partner's assistance is voluntary and should be respected and appreciated. Be very careful that you're not taking it for granted.

3. An awkward but insightful question: If your partner passed away or left you, would you continue farming? If the answer is no, perhaps you're honestly not a solo farmer in the first place. If the answer is affirmative, that you would absolutely continue farming, then you would figure out how to make it work. There are then things you can do currently in your solo scenario, such as hiring help or scaling down.

"When you decide if you could truly farm solo, it puts your partner's contribution into perspective," adds College. "When I realized that I would keep farming if I were here without Greg, then a great weight lifted. I recognized that the farm was my job and that I would find a way to get it done. While the occasional help would be appreciated and even less occasionally requested, it would never be expected."

Idea Seed:

"Eight years in, I've finally realized that the farm and the business will always be in a state of flux. Farming, it turns out, is not a linear march toward static success. Every season throws you some unexpected curveballs, no matter how many years you've been at it. I've learned a heap from those experiences, particularly the setbacks: A lifetime of learning from my mistakes might just turn me into a wise old farmer someday."

— Zoë Bradbury, Valley Flora Farm, Langolis, Oregon

While you're high in agriculture ambition, there's one green you probably lack: cash. You lack money toward a farm down payment and probably don't have the consistent W-2 track record to qualify for a bank loan. Add in hefty student loan debt and you're feeling financially overwhelmed before you plant any seeds of your own.

Time is on your side. While you understandably feel ready and raring to go, you honestly do have decades of farming seasons ahead of you. With this comes the advantage of being able to swerve and detour when a favorable circumstance serendipitously comes up.

Take advantage of the wealth of educational resources out there and develop your expertise and knowledge base. Embrace this free-as-a-bird life phase and dive into on-farm apprenticeships and experience farming a vineyard in Italy or harvesting marijuana in an intentional community in California. Tap into the growing numbers of beginning farmer programs and be sure to ask about any scholarships or volunteer options available to keep costs low.

Your commitment and dedication may be high, but the challenge is to keep that percolating over what might be a few years until you are at a point financially and experientially to start your farm. Often what starts as part-time work can take focus away from your farming vision. Or you might be offered a great job at a nonprofit doing food and agriculture related work. Whatever the situation, control the time this paid work takes, so that it doesn't take over your life to the point that the farm dream gets squeezed out.

Additionally, love and life happen in between everything else. Your fiancé starts graduate school, and you move with him to New York City. You meet your soul mate, who happens to be a single parent with two young kids, and your life quickly snowballs to full-scale family mode. In these cases, the life experiences fall into the happy category, but they still need to be managed in partnership with your farm goals. Tragedy can sometimes twist its way into our lives as well — the death of a loved one or an unexpected personal

health crisis. There will always be life elements competing for your focus and affecting your farm timeline and plans.

Maybe family life absorbs you to the point that farm dreams take the back burner. A three-year plan can evolve into a five-year plan, but if your farm dream keeps taking the back seat to other competing life priorities, it will never happen. Keep taking small steps — read books and attend conferences — to keep moving in a forward direction.

As fledgling young farmers, it can be hard to have your voice heard or taken seriously, even by old-timers in the organic community. "That will never work," is a response you might get to your vision of a permaculture

Idea Seed:

"Keep records of your farm training and experiences when they happen. One day you may need these as formal documentation when applying for a loan or other farm program. Many of your educational experiences may be unconventional and not fall into the normal 'receive a grade and transcript' format, so create your own formats. Keep a log of the conferences and workshops you attend, too. As you wrap up a farm apprenticeship, ask the farm manager or owner for a letter of reference. It's much easier to ask for such a reference when your experience is still fresh in everyone's mind, versus a few years down the road."

— Jan Joannides, executive director of Renewing the Countryside

operation or intensive urban farm operation. Don't despair or try to fight it. Your best recourse is to stay the course and keep committed to your vision. That said, find an ally.

In my local tribe in Wisconsin, when we gather regularly for informal potlucks, ages range from early twenties to nineties. Information, ideas, and inspiration respectfully flow like the wine. The key is finding your tribe, that group of women who can both respect your vision and provide feedback.

4. Family farm boomerangs: returning to family farmland

For you, "family farm" really means something. Maybe those words refer to the place you grew up on or another family member's farm with which you have strong ties, such as your grandparents' ranch. Maybe as a young adult you swore you would never return to this place. Years of life experience have left their mark, and you now yearn to return home. On a mission and perhaps on limited time (boomerangs can also be encore farmers), you are Committed with a capital C to this life chapter of yours. Your roots and ties run deep and strong. Returning to the farm with a career behind you, you may bring some

Idea Seed:

"The tenacity of women to do something for the good of their community, the environment, and future generations never ceases to amaze me. Being a woman farmer and growing food sustainably while engaging with community is certainly this."

— Temra Costa, author of
Farmer Jane: Women Changing the Way We Eat

savings plus managerial experience to get the job done.

Family roots can be complicated. Coming back home to the farm may involve your own vision for the land. You may be more sustainable and conservation-minded than your siblings or other relations, which can add a layer of negotiations. While your primary mission remains keeping your family farm alive and thriving, you may also have additional financial pressure to make the farm business profitable as you need the cash flow coming in.

"A best first step in creating productive family communications where everyone is heard and respected is to let everyone tell their stories, share memories and vision for the farm," advises Lynn Heuss, program coordinator for WFAN, who has worked extensively with the Women Caring for the Land project, supporting women in moving family farms toward conservation practices. "Understandably, you have your perspective of how life should go, but listening to others involved sets a tone of mutual respect. You may also be pleasantly surprised by how you and other family members are not that far apart in your vision, which is a much more fruitful place to start from."

Perhaps your family no longer farms the land but rents it to area farmers. This tenant relationship puts you in an interesting position as you can lead the rental process with your conservation values in mind. Ruth Rabinowitz learned this process as she took on managing the rental of her family's farmland in Iowa, though she calls Santa Cruz, California, her home.

"For over thirty years, my father operated passively when managing this land," explains Rabinowitz. "Like many non-operator landowners, he put his trust in the management company to take care of the rental relationships and do the right thing for the land." As Rabinowitz's father grew elderly, she came on board, asking questions and keeping the long-term health and conservation of the land first and foremost.

"I've learned these land rental relationships go beyond dollar signs," advises Rabinowitz. "I haven't always gone with the highest bidder, because sometimes that bidding farmer is just wanting to mine the soil and not add back nutrients. It's easy to be lured by cash, but the soil then pays the long-term price."

Rabinowitz also recommends developing strong personal relationships with your tenant. "Think of the owner and lessee as a team. You want a farmer on your side who shares your passion for leaving the land better than we found it," offers Rabinowitz. New herself to agriculture, Rabinowitz dove into educating herself on everything her tenant would need to know, from understanding how to read soil maps to developing knowledge of soil health. "I read everything I can get my hands on, go to conferences and workshops, and ask questions constantly. I found that my commitment to learning and understanding land management went a long way in earning the respect from the farmers renting our land as they knew I wanted healthy soil to help their business succeed long term."

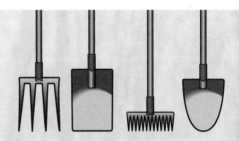

Tool Shed:

Resources for Managing Family Farms

Women Caring for the Land
www.womencaringfortheland.org
Free conservation training specifically for women landowners, facilitated by the Women, Food and Agriculture Network (WFAN). These workshops are for women only, providing situations for women landowners to connect with area female agency staff on potential programs and resources for land management.

Farm Transitions
www.farmtransitions.org
A collaborative venture among various non-profits, Farm Transitions provides free tools and templates for family farms to craft long-term management plans that address multiple goals and needs.

Drake University's Sustainable Agriculture Land Tenure Initiative
www.sustainablefarmlease.org
This program offers a variety of free tools to support landowners and farmers to develop farm lease arrangements that are profitable and sustainable for the landowner, the farmer, the community, and the land.

How She Sows it:

Relinda Walker, Walker Organic Farm, Sylvania, Georgia

How to Cultivate New Visions on Family Land

"I guess I've always been a little different," shares Relinda Walker with a warm smile. In her case of pioneering organics on her family's farmland in Georgia, different is not only good, it's a strategy that cultivates business success along with health and longevity of her family land.

Walker grew up on what today forms Walker Organic Farm. "I had limited involvement on the farm growing up. My dad, who primarily grew traditional row crops, ran the operation successfully and was among one of the first in our area to raise cantaloupes and watermelons commercially."

As with many farm kids, Walker left for college and moved on to a successful business career in manufacturing and technology for several decades. Situations in business and life change, however. "I saw the writing on the corporate cubicle wall that the manufacturing climate was changing and I, too, was ready for a personal change. After 9/11 happened in 2001, I took a voluntary layoff later that year.

Relinda Walker in the field at Walker Organic Farm.
JOHN D. IVANKO

I had been mulling over what was next for me. I went to hear Joan Gussow, author of *This Organic Life*, speak, and then it all came together for me; a vision clicked. All roads now led for me to come back home and grow food."

Walker Organic Farm launched in 2005 as one of the first organic operations in southern Georgia. It raises specialty varieties of vegetables, including signature crops such as rainbow carrots in vivid shades of purple, orange, and red; onion seedlings; and various cool-weather crops such as collard greens and kale. "Chefs were our key initial market, and we were the first to offer unique items for high-end restaurants." Currently, most sales come from local farmers' markets, followed by restaurant sales and local resellers.

As a solo farm woman returning to farmland with an organic, local-food mission, Walker offers this advice to others contemplating cultivating similar paths:

Relinda Walker of Walker Organic Farm, Sylvania, Georgia.
JOHN D. IVANKO

Tap agency resources

"Start by walking into your FSA [Farm Service Agency], Extension, or any other state or federal office you can find, and ask them what resources are available for you and your farm," advises Walker. In some cases, particularly if you're one of the first organic ventures in your area, you may need to push a bit and help educate agency staff not familiar with your needs and practices. Share organic certification fact sheets and examples of what you can and can't use for pest control. Georgia Organics has an entire free "Fundamentals of Organic Farming and Gardening: An Instructor's Guide" online, with various fact sheets and information.

"Eventually you will find a kindred spirit of support. For example, my Extension agent admitted he didn't know anything about organics but would learn along with me. That's been a strong partnership ever since."

Grow farmer connections

"Support and advice from other farmers in sustainable agriculture are key to both your education and sanity," Walker shares. Don't worry if there isn't a large and thriving community of farmers locally; these networks will grow with time and your nurturing. "A potluck group may start with just you and one other farmer; don't worry, and keep meeting — others will find you."

Walker co-founded the Coastal Organic Growers, a regional group of organic farmers that still gets together for potlucks. She also regularly hosts field days and tours for agency staff, area farmers, and the general public. "There's nothing like being able to bring people to your land and let them taste something from the fields to connect folks to the land and their food source."

Prioritize Cash Flow

"While you, like me, may have a strong mission that drives what you do on your family land, always remember that the farm is a business that needs positive cash flow to survive," offers Walker. She tapped into FSA for a debt consolidation loan to help keep the local people she hired on the payroll while she worked toward the long-term financial health of the farm.

Walker's return home blended both this new farm business venture with the need to care for her elderly father. "This proved to be a rewarding new phase of life . . . as he could share his knowledge and farming wisdom," adds Walker. "At first he couldn't understand why we were growing unique crops like baby vegetables, but then he saw and appreciated the growing market."

What's next for this spry, energetic woman as she enters her seventies in the next few years? Walker, a single woman with no kids, is exploring possibilities to keep Walker Organic Farm vibrant and viable into future generations. One option might be to create an incubator farm, where young farmers could come and lease land affordably and access infrastructure, tools, and Walker's years of experience to help launch their own ventures. "I'm starting to shift more of the physical farm labor to my staff. I keep fit by walking the land but am frankly spending more time than I wish in front of the computer, managing communications and running farm errands."

"I wouldn't recommend farming as a retirement strategy," laughs Walker. "But I wouldn't want it any other way. There's nothing more rewarding that I've done in my life than raising good food."

In this "Understanding Our Roots" kick-off section of *Soil Sisters*, we dug into our past. By understanding both our history as women growing food as well as our personal stories as we transition into farming, we create fertile ground from which to grow. You won't be navigating this journey alone. You come to the field rooted in a deep history of women growing food along with support from that growing group of females farming today. Time to get growing, sister!

PART 2:
GLEANING KNOWLEDGE

Farm smarts

- Ask the same question to multiple people.
- Remember that there is no right path.
- Maintain a log of all the related books you read.
- Study and read, but remember that Mother Nature decides.
- Keep a winter reading list.
- Play in the dirt.
- Recognize that everyone you meet knows something you do not.
- Play by your own rules.
- Don't let your mistakes define you.
- Learn from new and unconventional angles.
- Embrace life-long learning.

Chapter 3

Organic Trends

THE FIRST THING THAT HAPPENS IN THE MORNING around here is a weather check. Look at the Doppler radar online and read the forecast. See if a cold or warm front is coming in or out. We put that info together and make our best guess on how to plan the workday. If it's a clear spring morning with rain rolling in late afternoon, plant the last of the mesclun ASAP. In the summer when we see that inevitable string of consecutive sweltering hot days, we finish up outside chores at the cool of dawn.

We take in this weather information and make the best decisions we can. Mr. Weatherman Al Roker himself doesn't nail all his forecasts; nor do I from my laptop. A spontaneous July thunderstorm erupts, giving the laundry on the line an extra rinse. The lettuce greens just sprouting get a real beating. But most of the time, by assessing different perspectives and pieces of weather information, we can plan what we need to get done and determine how to best to get there.

The same approach works with figuring out how you fit into the bigger food and farming picture. What will business interest be like a year or five years from now? Or better yet, how can you tap into what's hot and upcoming?

This chapter is your Doppler on food and farm business trends, identifying some key "fronts" — some emerging topics that will most likely continue to increase in interest and create a strong marketing foundation for your venture.

Certifiably organic

Congratulations! You walked into this food and farming party at the perfect time. From shoppers wanting to know the farmer's face behind their food to

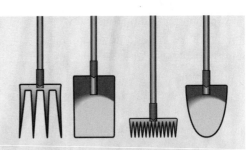

Tool Shed:

Take on the Terms

As the organic food movement grows, so does its vocabulary. Check out this primer on key words you'll encounter. Some have official legal definitions, and others have generally accepted definitions that can often be manipulated in the marketing game.

Certified organic

In 2002, the USDA issued organic certification standards that outline practices and procedures required for farmers to sell their products as organic and food processors to use the green-and-white organic label. Certification involves a trained inspector examining the farm and a certifying agency reviewing the farmer's application and inspection report. Overseen by the National Organic Standards Board, these standards evolve and (hopefully) improve based on feedback, including farmer input.

Transitional

For a first-time farm going through the certification process, the soil must be free from use of prohibited substances for several years before it can be certified organic. This is known as being in transition.

Sustainable agriculture

Though there is no single definition and this term is not regulated, it generally refers to farming practices that protect and prioritize the environment, public health, human communities, and animal welfare.

Biodynamic agriculture

A "spirit meets ethic meets ecology" approach to agriculture, biodynamics goes beyond organic standards to include a very holistic and cooperative approach to health, including preparations made from fermented manure, minerals, and herbs to restore soil nutrition. Certified biodynamic farms need to first be officially organic certified.

GMO

Meaning "genetically modified organism," a GMO is an organism whose cellular make-up has been forced through genetic engineering so that its DNA contains one or more genes of an entirely unrelated species, such as fish genes pumped into strawberries to apparently "protect the fruit from freezing." Seriously. Dairy cows get injected with the genetically engineered hormone rBGH (recombinant bovine growth hormone) to increase milk production. Certified organic products do not contain GMOs. ☞

Non-GMO project

This is North America's only independent verification system for products made according to "best practices for GMO avoidance." Retailers started this as an option for consumers wanting clearly labeled non-GMO food and products.

Certified naturally grown (CNG)

This is a farmer-initiated alternative to organic certification specifically tailored for small-scale, diversified direct-to-market producers. Volunteer neighboring farms serve as inspectors.

Locavore

Appearing in the *Urban Dictionary* in 2007, "locavore" describes someone who intentionally eats food grown or produced locally or within a certain radius such as 50, 100, or 150 miles, usually for ecological reasons.

Industrialized, conventional, or chemical agriculture

This is a catchall term for larger-scale farming practices that use pesticides, fungicides, and herbicides as a regular part of their farm management plan. They also focus more on the plant than the soil nutrients and fertilizers must be applied to maintain crop yields.

Grass fed

Animals on managed pasture, typically beef and lamb, are described as grass fed. When they are indoors, as during the winter, for example, they're fed hay or an equivalent to hay, but never grain.

Pasture raised

Animals such as pigs and chickens cannot exist exclusively on a diet of grass. These animals are on fresh pasture but also receive some grain supplements.

travelers increasingly seeking on-farm experiences, there are multiple opportunities for your farm business to tap into.

"The moment is ripe for women food entrepreneurs because of the many great resources available in support of local food, but also because of the steadily increasing consumer demand for certified organic ingredients and products," proclaims Carla Wright, executive director at the Organic Processing Institute, a nonprofit organization championing organics and supporting food entrepreneurs with resources and training. "There is no better time to develop a business that will position you as an organic food entrepreneur. This is true of all sizes of enterprises."

No surprise, women aren't the only ones wanting to get their piece of the organic profit pie. As big corporations buy up smaller organic brands, it becomes increasingly difficult for the public to know where their money is going. Products that appear to come from a small eco-enterprise may actually belong to big agribusiness guns like Coca-Cola, which owns a stake in Green Mountain Coffee, or Mars Inc., which owns Seeds of Change. Since 1995, the number of independent organic food processing companies plummeted from 81 to 15, according to the Cornucopia Institute. This decline occurred while the organic industry continues to grow at an average of ten percent annually. The result is that most of the commercial organic food on our shelves is produced by a small group of large international corporations like Campbell's and Kellogg's.

It's so ironic you almost have to laugh — or cry. These mega-corporations with organic brands use their profits for lobbying influence to push agendas that directly conflict with the good-food movement. For example, Coca-Cola, owner of Honest Tea and Odwalla, contributed over one million dollars to oppose California's initiative to mandate GMO labeling. Money talked as this influx of corporate dollars turned the tide to defeat California's Proposition 37 in 2012, which would have required labeling of most genetically engineered food.

Trending now

Female farmers and food entrepreneurs create our own playing field, launching small-scale, community-focused ventures that reflect our shared values of sustainability and leaving this world a better place. Take a look at four key food and farm trends and start thinking about how this fits into your business plan. If you're at the beginning, brainstorming stage — better yet. You can craft your vision to fit into these timely hot topics:

1. Organic food

Total organic sales in the United States continue to rise, up nearly ten percent annually and growing from $1 billion in 1990 to more than $31.5 billion in 2013, according to the Organic Trade Association (OTA). Organic food sales in 2014 in the US totaled $35.9 billion, up 11 percent and setting a new record.

This growth in demand for organic food is largely driven by parents aiming to make healthy choices for their families. According to the OTA, a 2014 survey of over 1,200 households across the US with at least one child under 18, found that nearly 25 percent of parents already buying organic said that avoiding GMOs is a top reason they choose organic. Only 16 percent of parents said the same in 2013.

"Parents in charge of the household budget recognize the benefits of organic, and they're willing to pay a little more to know that they are giving their families the highest-quality and most healthy products being offered in their local store," says Laura Batcha, OTA's CEO and executive director. "Parents have become more informed about the benefits of organic, and they have also become more aware of the questions surrounding GMOs. That heightened awareness is being reflected in their buying decisions."

These questions stem from the fact that there still are a lot of unknowns regarding potential long-term negative effects of GMOs. Parents increasingly embrace a precautionary principle: If we're not sure whether GMOs have consequences, let's keep them out of our kitchens and bodies until more research is done.

This same approach is taken by the European Union, which has some of the most stringent GMO regulations in the world. All GMOs, along with irradiated food, are classified as "new food" and are thereby subject to an extensive, case-by-case, science-based food evaluation by the European Food Safety Authority (EFSA). Not surprisingly, the EFSA has approved only a handful of GMOs. Our federal government, on the other hand, keeps blindly adding GMOs to our food system. As the parents who are driving the organic market realize, one way to guarantee that your food doesn't contain GMOs is to buy certified organic.

Moms, in particular, play an important role in growing this demand for organic foods, especially as research increasingly supports the positive effects of an organic diet on children. "Children encounter pesticides daily and have unique susceptibilities to their potential toxicity," the American Academy of Pediatrics wrote in 2012, and "chronic health implications from both acute and chronic exposure are emerging."

2. Local food and farmer connections

Move over Frosted Flakes. Meet Farmer Fran. Though industrialized agriculture and corporate food agri-businesses still dominate supermarket shelves, shoppers increasingly detour out of those aisles and down the street to buy direct from local farmers at farmers' markets. In 2014, there were 8,268 registered farmers' markets across the nation, an increase of nearly 50 percent from ten years earlier, according to the USDA Agricultural Marketing Service. The USDA's "Know Your Farmer, Know Your Food" campaign also aims to connect growers with customers.

Faye Jones harvests greens on her farm.

MARK PLUNKETT

Faye Jones, Midwest Organic and Sustainable Education Service (MOSES), Spring Valley, Wisconsin

Growing the Organic Market, One Farmer at a Time

"Right now is the best time ever for women interested in farming and passionate about organic and sustainable agriculture to explore launching an operation," shares Faye Jones, executive director of the Midwest Organic and Sustainable Education Service (MOSES). "Between the good-food market momentum and the long menu of training and educational options out there, plus increasing opportunities to connect with and learn from and support other women farmers, you can readily set yourself up for success in a way that wasn't possible when I started farming three decades ago."

Jones first connected with organic farming at 19 when she visited her boyfriend's family's dairy farm. "It was a wonderful, magical experience with cows, pastures, barns," she says, beaming. Inspired, Jones enrolled in agricultural courses but soon grew disappointed as they focused only on conventional farming with heavy mechanical input and toxic chemicals. Jones dropped out of college for her own "learning by doing" curriculum that included farm apprenticeships, learning from mentors, and attending conferences. "The organic movement was such a young baby back then that when I use the word 'conferences' they were nothing like what you picture today. In the 1980s, these were more like reasons to get together and share knowledge. We were all in the same boat, learning together as we farmed and collaboratively sharing together."

In the early 1980s, Jones started farming vegetables

and flowers on rented parcels that were part of existing organic farms selling to the Minneapolis area. "This worked out great because I had mentors on site that I could keep learning from." In 1989, Jones purchased her farm in west central Wisconsin and started moving toward specializing in cut flowers. "Twenty years ago, the timing was ideal to get into flowers, as you started seeing flowers in places other than a flower shop. Upscale grocery stores sold bouquets that featured blooms other than carnations and roses, and people began viewing flowers more as a gift to give any time, rather than just for Valentine's Day or when someone is sick," she recalls.

"Also, when I did my books at the end of the year, I quickly realized I made a lot more profit off the flowers for the time I invested versus the vegetables."

Still, as a beginning farmer in this new flower market, Jones knew she needed to amp up her marketing and customer service mojo to create a viable sales base. "I approached food cooperatives in the Twin Cities and made it super easy for them to say yes and sell my flowers. I'd bring the bouquets directly to the store in plastic sleeves, labeled and ready to sell from clean black plastic buckets I provided."

Jones wisely also offered the bouquets on consignment for the first two weeks to get her foot in the door. "They couldn't lose with me because for those first couple of deliveries I'd take back anything they didn't sell," she adds. Business grew, and within a few years she needed to turn down accounts as she still wanted to run Morning Glory Farm as a solo operation and have time for her daughter, Nina, whom she was raising as a single mom.

Meanwhile, Jones kept learning. She kept going back to that organic farming conference in the Midwest, which was now growing from the original 90 attendees who attended in 1989. "I was shocked that an organic conference like this didn't serve organic food," remembers Jones. Not one to just vent and leave, Jones took on organizing the food, which opened up a path she never imagined. "Yes, the food tasted better, but what happened for me personally is I realized I had a knack for organizational leadership. I've always had this life dream of wanting to help change what farming is all about in America, and working with a nonprofit gave me the platform to do this."

The next chapter of Jones's life started calling, as she became more involved with MOSES. She became the first paid staff at MOSES 17 years ago and today still leads the organization. In 2005 she traded farming for full-time work at this nonprofit. MOSES runs the organic conference, which serves over 3,500 enthusiastic attendees. "We like to call it the largest organic gathering in the known galaxy," she says with a grin. And the organic food is great, too.

"Identify what you need personally in your life and match that with the right situation," advises Jones. "Like my experience with getting into the flower market early on, keep your eye on what areas might be untapped right now." Jones advises against blindly following established market routes. "While farmers' markets make up core sales outlets for many small-scale growers, do your research before committing to one because increasingly metro areas may be saturated with too many markets too close to each other." Or think about alternatives such as sales to restaurants or schools.

Growing the organic and sustainable agriculture movement involves a lot of different roles to play and jobs to get done. "At different stages of your life, be open to ideas outside of the farm field as a way to have an impact changing our food system. We need women in all positions of agriculture, from the tractor seat to board chairs and everything in between."

Research from Sullivan Higdon & Sink's Food Think, "A Fresh Look at Organic and Local" in 2012 found that 70 percent of consumers would like to know more about where food comes from. "The vast majority of consumers (79%) would like to buy more local food, and almost 6 in 10 (59%) consumers say it's important when buying food that it be locally sourced, grown or made."

Buying local also supports a healthy community economy. According to the Institute for Self Reliance, in a comparison study of local and national chain retailers, local stores return a total of 52 percent revenue to the local economy, but it's only 14 percent for the chain guys. Similarly, local independent restaurants recirculate an average of 79 percent of their revenue locally, compared to 30 percent for chain eateries.

3. Specialty foods

Love your aged green peppercorn brick cheese, pickled beans, and blueberry ginger flavored kombucha? You're not alone. These types of products fall under "specialty foods," a booming industry experiencing 22 percent growth between 2010 and 2012 and topping $85 billion, according to the Specialty Food Association.

Two of the most likely characteristics of new specialty food products coming onto market are gluten free (38%) and convenient/easy to prepare items (37%), according to the Specialty Food Association. An easy way to break into specialty foods is small-scale, value-added processing in your farm kitchen under cottage food laws, state legislation that authorizes specific non-hazardous food items made in home kitchens for sale at certain public venues. More on this in Chapter 5. Cottage food businesses are covered in detail in my book *Homemade for Sale: How to Set Up and Market a Food Business from Your Home Kitchen.*

Don't let the epicurean sheen of specialty foods lead you to think it's high-end foodie fluff. Many possibilities exist to create a specialty food business that exemplifies your values and can be a platform for your educational mission.

"This isn't just boutique or gourmet food artistry," says Jordan Champagne, who with her husband, Todd, started Happy Girl Kitchen in 2002, now based in Pacific Grove, California. "Preserving the local harvest is about food security and rediscovering how to eat local year round. My love for preserving the harvest came to life when Todd and I lived for a summer on a family fjord farm in Norway, tending the garden, milking the cow, and preserving food," she reminisces. "Over there, I realized that if you wanted to eat the fruits of the short summer year round, you needed to preserve them for the long winter."

After that overseas experience, the Champagnes launched Fearless Pickles back home along the central coast of California. They did it all: raised the cucumbers, pickled and packaged them, and coordinated sales. "It quickly became too much to do, and I realized we're surrounded by such amazing farmers growing in abundance that it's much easier and more efficient to buy direct from them, especially women growers like my dear friend, Jamie Collins at Serendipity Farms."

Champagne reinvented the business into what is now Happy Girl Kitchen and Café, specializing in crafting value-added products in jars, using local California produce purchased directly from farmers. She's also keen on using imperfect produce that might otherwise end up in compost bin or tossed to the farm animals.

"Think of the kitchen as a place to play, experiment, and have fun," Champagne advises fledgling farmers and specialty food entrepreneurs. "Pair unusual flavors together and see what happens." That's solid advice coming from Champagne, as the sweet and savory combination in her apricot chili jam

Jordan Champagne of Happy Girl Kitchen, alongside her daughter, packing lemons in crocks.

JOHN D. IVANKO

scored a Good Food Award, dubbed the "Oscars of the food movement."

Part kitchen, part retail store, and part café, Happy Girl Kitchen and Café creates a community spot for those seeking these food connections. Champagne warmly welcomes and chats with customers in this lively 2,700-square-foot space where she and her staff pump out dozens of products as well as hosting preservation workshops and pop-up dinners. "It's about building a community around your food passions," she sums up. "That's the cornerstone of any successful food business committed to sustainable and organic agriculture."

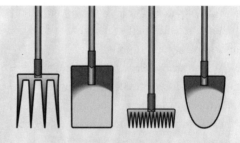

Tool Shed:

The Good Food Awards

Draw ideas from award-winning masters: Study the list of Good Food Awards and harvest ideas. Are certain food categories offered in some areas of the country but not yours? What can you learn from the packaging, websites, and marketing approaches? You'll quickly see that the vast majority of these companies, like Happy Girl Kitchen, build brand through authenticity — sharing honest stories about what motivated the upstart and where the ingredients are grown.

Launched in 2010, the Good Food Awards came to life through a collaboration of food producers, farmers, food journalists, and independent grocers aiming to recognize "truly good food," the kind that brings people together and builds strong, healthy communities. Winners are honored at a special ceremony in San Francisco in January.

www.goodfoodawards.org

4. Agritourism

Two things we soil sisters love: our farms and sharing that connection to the land and food with others. Make agritourism part of your business plan and you've hit the industry trend jackpot. On-farm experiences are hot in demand, particularly farm-stay B&B experiences and farm-to-table meal service.

Many city folks love to pay for the "play farm" experience, and savvy farmers are stepping up to the demand. From 2007 to 2012 the number of US farms catering to city folk went up by 42 percent, bringing in more than $700 million, according to the latest Census of Agriculture. Since 2007, the amount of money brought in under agritourism rose by 24 percent.

Travelers increasingly seek authentic, slice-of-life experiences. If that slice happens to be pie served on a farm and made with homegrown rhubarb, all the better. In a study by *National Geographic Traveler* and the Travel Industry Association of America, 55.1 million US travelers are classified as "geo-tourists," travelers who are interested in nature, culture, and heritage tourism. Similarly, according to The International Ecotourism Society (TIES), about 17 million US travelers consider environmental factors when patronizing businesses, and about half prefer trips with small-scale accommodations operated by locals.

"Farm stays and female farmers go together like the homegrown bacon and eggs she'll be serving," explains Jan Joannides, co-founder and executive director of Renewing the Countryside (RTC), a nonprofit dedicated to championing and supporting rural and farm-based businesses. "Farm stays showcase our natural female knack for hosting with warm hospitality, which makes an on-farm lodging experience a natural fit for women farmers looking to get into agritourism."

"I'd say 90 to 95 percent of the farm stays run in this country are operated by women," shares Scottie Jones, founder of Farm Stay U.S., a national portal connecting travelers to farm experiences and supporting farm-stay start-ups with resources and education. "The farm may be co-owned by a partner or spouse, but the job of running the farm-stay operation is taken on by the woman, as hospitality falls more naturally to women, even in this day and age."

Jones estimates that about 1,000 working farms, ranches, and vineyards in the US offer lodging. She runs Leaping Lamb Farm in Alsea, Oregon, a lamb operation that also offers a farm-stay cottage rental. Jones founded Farm Stay U.S. to help make these experiences more mainstream travel options, like they are throughout Europe, Australia, and other parts of the world.

Increasingly, these travelers also want more than just a farmstead bed. They want an on-farm dinner too, opening up a wide menu of opportunity to include some form of food service in your operation. Over the past ten years, the role food plays in the travel industry has grown tremendously, fueled by everything from *Food Paradise* and *No Reservations* shows on the Travel Channel to the swarm of tempting food photos invading social media feeds. According to the World Food Travel Association, travelers increasingly seek out the unique and the flavorful, with dining consistently rated as one of the top three favorite tourist activities. These travel and food enthusiasts are particularly seeking authentic agricultural experiences and "insider" perspectives, something a farmer and farm-direct dining experience can offer.

On-farm meals dovetail with the 2015 Culinary Forecast by the National Restaurant Association in over half of their top ten food trends, including the top three:

1. Locally sourced meat and seafood
2. Locally grown produce
3. Environmental sustainability

Further down the top ten list and also applicable to women farmers are

5. Natural ingredients/minimally processed food
7. Hyper-local sourcing (for example, restaurants making artisanal items in-house, such as ketchup, pickles, and cured meats).
10. Farm/estate-branded items

Local markets with moxie

As you decide the direction of your farm livelihood, perhaps you also have flexibility as to where you land geographically. While some of us may already own property or come with regional ties to a specific area, others may be open to other parts of the country.

With farmland prices increasing (doubling from 2000 to 2010, according to the USDA) you'll want to be strategic in where you go so that you can afford the acreage and still have access to a larger urban market if needed for sales.

"There still are some US cities that offer both affordable farmland within driving distance and that haven't been oversaturated with small-scale producers and CSAs," shares Rebecca Thistlethwaite, a farming consultant and author of *Farms with a Future: Creating and Growing a Sustainable Farm Business* and the new book, *The New Livestock Farmers: The Business of Raising and Selling Ethical Meat.* "Look beyond the known foodie hot spots like Madison, Wisconsin, where some markets have a five-year wait for a farmers' market booth, and be a part of a smaller but growing local food market that will more readily embrace what you have to offer."

College towns have a well-educated populace that are often seeking better, fresher sources of food and have incomes to support that, explains Thistlethwaite, who determined her list of the eight hotspots based on her travels and book research:

> Lawrence, Kansas
> Mobile, Alabama
> Moscow, Idaho
> Gainesville, Florida
> Lincoln, Nebraska
> Grand Junction, Colorado
> Fayetteville, Arkansas
> St. Louis, Missouri

"Aside from St. Louis, these are also towns under 200,000 people, where I think it is easier to build community and strong social networks, which is essential to building a solid business." Likewise, most of these places still have a strong agricultural heritage, which makes it easier to find things like tractor mechanics, equipment rentals, grain mills, slaughter houses, and other agricultural infrastructure.

"Remember, there are many places in this country where land is more affordable and the regional populace is supportive of diversified market gardeners and pasture-based animal producers," Thistlethwaite adds. "We need to get away from our obsession with the two coasts and start feeding the rest of the country good food."

"Understandably, hip food towns like Asheville, North Carolina, or Portland, Oregon, or the San Francisco Bay Area have appeal, and it's alluring to be in a place where there's a lot of people like yourself," adds Thistlethwaite. "But if you're willing to move to an area where the local food movement is more nascent or less known, you could ride that early wave and both create a strong business as well as help contribute directly to the community leadership." As in any business, it helps to be an early adopter, carve out your market niche, build

Idea Seed:

Don't believe the hype; get your own data.

Intuit, trust, and verify.

Notice emergent possibilities; kindle them gently.

Build fierce loyalty with people who know how to motivate themselves.

Inspire them and gift them a lot of freedom.

— Severine von Tscharner Fleming, director of Greenhorns, co-founder of Farm Hack, National Young Farmers Coalition

customer loyalty, and maintain it over the long term. "Coming in later after the market is thoroughly saturated can be tough," she sums up.

We'll go into more detail on specific business ideas and how to start them in Chapters 5 and 6, but for now, consider this a foundation of ideas on which to build your farm livelihood. Like the weather forecast, life and business don't come with guarantees. We can, however, research and assess likelihoods. Celebrate the good news that you're coming to the food and farming scene as the movement is rolling. Enjoy the ride!

Chapter 4

Key Ingredients for Success

IMAGINE WALKING THROUGH ROWS OF LUXURIANT TOMATO PLANTS, the branches almost bending to the ground with abundant, ruby-red fruit. Everywhere you look beyond the tomatoes, you see something to harvest. To your right sits a long row of prolific cucumbers. To your left you see zucchinis likely to explode if you don't clip them right now.

This summer harvest scene is akin to the abundant possibilities as you launch into a farm business. Never has there been a more plentiful time for a beginning farmer, particularly a female grower, to learn, plan, and launch. Thanks to the snowballing growth of sustainable and organic agriculture, an increasing number of resources exist, including nonprofits, universities, and federal agencies.

But like walking through that fertile field, too many options can feel overwhelming. Where to start? Pick the tomatoes or that Eight Ball zucchini about to detonate into a bowling ball? Should you commit to a full-year "Farm Beginnings" course to develop your business plan or string together on-farm apprenticeships or self-teach through books and online courses?

Gleaning refers to collecting leftover crops, heading into the field after the main harvest when much goodness still remains. This concept also means finding things out slowly, gathering up something bit by bit. Since there's so much out there, you will never lack knowledge to harvest. Take things inch by inch, row by row. You'll acquire the skills, contacts, and confidence to successfully instigate your venture. Dive into the details of creating a solid base of farm smarts, from understanding food trends and USDA acronyms to tapping into training programs and funding possibilities.

Consider this chapter a roadmap not a blueprint. This is a case of multiple approaches to the same endpoint. Some of us may take the tollway and arrive at the destination rapidly; perhaps by entering an intense training program, buying land, and starting farming the following season. Others may wiggle more slowly along the back roads, making detours and refining our vision along the way. That path may entail attending conferences, apprenticing at a couple farms, meeting some beginning farmers with the same vision, or forming a farm cooperative and renting land from a trust. Same destination but different paths.

This chapter aims to give those new to the agriculture world a quick primer on the possibilities. We'll look at three key ingredients to building your farm: knowledge base, access to land, and cash crop.

Build your knowledge base: what you need to know

Most farmers will gladly give you an opinion on something, but I doubt you will ever meet a farmer who proclaims to be "done" learning about agriculture. Whatever our years of experience, we always add to our knowledge base. A big reason for this is, frankly, we're challenged and fascinated by the watermelons in the field or pigs in the pen. How can we do something better next season? How could we have better dealt with drought, blight, or any other problem we experienced this past year?

"When a beginning woman farmer asks me what's the best route to learn, I say take them all," explains Harriet Behar with a grin. Behar is the organic specialist on staff at MOSES, and she fields thousands of questions from organic farmers along a wide continuum of experience. "Seriously, the best thing you can do is a hybrid approach. Go to conferences and workshops, read books and fact sheets, talk to other growers, come out to farm tours. Most times in farming, there isn't one specific right way to do anything, so exposing yourself to multiple perspectives strengthens your knowledge base."

Are you someone who learns best independently, creating your own curriculum and learning mode? Or do you prefer a more structured setting, perhaps in a classroom with other people? Both options are viable; just place yourself in a situation where you thrive. Unless it's an online program, structured options typically require in-person time in lectures, workshops, and other learning formats — which may require travel on your part. A perk of these programs is connecting with other new farmers on the same start-up road as you.

Operating independently, you can glean from a variety of resources at your own pace, setting your own schedule and priorities. Especially if you're juggling other jobs and family commitments, slow and steady may be the ticket. The downside can be that it's easy to get stuck in perpetual research mode. There will forever be another workshop to attend or book to read.

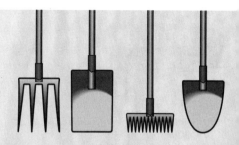

Tool Shed:

Women Farmers Starting with Structure

Cultivate your knowledge and connect with other women in these female-focused farming programs:

Annie's Project

extension.iastate.edu/annie/
Consisting of six education sessions, "Annie's Project" focuses on business planning and financial information, with a mission to empower farm women to be better business partners through networks and by managing and organizing critical information. Based out of Iowa Extension, "Annie's Project" offers programs in over 30 states.

Holistic Management International

holisticmanagement.org
"Empowering Beginning Women Farmers through Holistic Management Whole Farm Planning" takes an inclusive decision-making approach to planning one's farm.

Harvesting Our Potential, Women, Food & Agriculture Network (WFAN)

wfan.org
For women in Iowa and Nebraska, this program offers an eight-week on-farm internship and mentorship along with structured networking sessions.

North Carolina Women of the Land Agricultural Network

ncwolan.org
"The Farm School for Women" offers on-farm, hands-on training for varying lengths of time.

Farm Beginnings

farmbeginnings.org
Operated through the Land Stewardship Project in Minnesota, "Farm Beginnings" programs offer a structured year-long program of weekend seminars, field trips, and hands-on training. Not female-exclusive, "Farm Beginnings" works well for young farming couples aspiring to farm together.

To help you navigate your way through farm learning opportunities, here are five core resource and information access points:

1. United States Department of Agriculture (USDA)

Cue the acronyms; it's time to navigate the USDA. The USDA is that mammoth federal department that encompasses everything under the umbrella of farming, food safety, nutrition, and rural development. The USDA runs off funding and program priorities identified in the Farm Bill, a comprehensive omnibus bill passed about every five years that serves as the primary agricultural and food policy tool of the federal government.

Founded in 1862 under Abraham Lincoln's administration, the USDA was described as the "people's department," coming into existence at a time when the US economy was largely agrarian. Over the past 150 years we've detoured from our agrarian roots, shifting from being a country where half of Americans lived on a farm to a nation where we have twice as many prisoners as farmers. The USDA has justifiably earned a reputation as bureaucratic, benefiting large industrial farms and being bigger than it needs to be.

That said, I recommend that, as a beginning farmer, you start a relationship with the USDA. Although I can personally attest to the layers of administration and paperwork and resulting waste of time and resources, I also have met and worked with some of the most dedicated public servants committed to helping the next generation of farmers succeed.

For two years, I served as the panel manager for the Beginning Farmer and Rancher Development Program (BFRDP), a Farm Bill program supporting nonprofit and academic institutions that provide beginning farmer training. My role as an objective person outside the USDA was to help ensure that the grant allocation process is fair; one of my roles was to recruit beginning farmers to serve as grant reviewers. My colleagues at the USDA strongly supported the intent of the program being "peer reviewed," meaning that a group of farming peers decides how this pool of 18 million tax dollars was spent. I was able to bring twenty-something farmers into a Washington DC boardroom making these decisions. I'll be first to agree that the USDA has issues, but I'll also raise my hand in support that there are good people essaying to change things.

Here's a synthesis of major USDA departments you might cross paths with. Check out the USDA's new synthesized outlet for beginning farmer information: www.start2farm.gov

National Center for Appropriate Technology (NCAT):

www.ncat.org

With a mission to "[Help] people by championing small-scale, local, and sustainable solutions to reduce poverty, promote healthy communities, and protect natural resources," NCAT runs over 50 projects that support sustainability. Operating as a nonprofit, NCAT receives funding to manage various USDA programs under the conservation umbrella, including farm-to-school initiatives and the Applied Technology Transfer to Rural Areas (ATTRA), which has over 200 free publications, webinars, and staff on call to answer questions.

Sustainable Agriculture Research and Education (SARE):

www.sare.org

This is the USDA division that focuses on sustainable and organic information, offering free resources and publications and grant opportunities. Organized in four geographic regions, SARE aims to provide education and resources specific to each region's growing climate and needs. SARE is the most accessible portal within the USDA that directly funds farmer-led projects. (More on this in the section on funding.)

Farm Service Agency (FSA)

www.fsa.usda.gov

Nearly every county in the country has an FSA office with full-time staff. Think of the FSA as the finance arm of the USDA as it administers everything from farm loans and credit programs to crop insurance and emergency assistance for farmers and ranchers. (More on this in the section on funding.)

Natural Resources Conservation Service (NRCS)

www.nrcs.usda.gov

NRCS provides technical assistance to farmers and other private landowners, connecting you to various land management programs and incentives.

A challenge when working with both FSA and NRCS is that the majority of their programs support larger scale, commodity-based farms. Your local technical staff, most likely male, may not be familiar with programs like the Environmental Quality Incentives Program (EQUIP) or the Conservation Stewardship Program (CSP) or the needs of organic farmers, and it may test your patience and require some education to nudge them into a supportive role. Or you may find, like I did, that micro-sized farms like our 5.5 acres are too small to even qualify for a program like CSP.

An interesting ally at the FSA office may be the women who actually run the place: the administrators and secretaries. Men are then the actual FSA and NRCS agents. As this system slowly evolves, it can work to your advantage to befriend the women who will probably be the ones answering your calls or the first face you'll see when you walk in the door.

2. University Extension

www.extension.org

If you hail from the city, university Extension may not be on your radar, as it traditionally focused on rural communities. Your state land-grant university will run your local Extension, which is typically broken down by staff at the county level. You probably already know your state's land-grant institution. It's most likely your state's largest academic entity, such as Washington State University or Penn State University.

Publicly funded land-grant schools were started with the official intent to support agriculture education and research. The Morrill Act of 1862, sponsored by Vermont Congressman Justin Smith Morrill, gave each state 30,000 acres to sell and use the proceeds toward "agriculture and mechanical" education. Some states used the funds to start their own universities, like the University of Illinois, and some gave the money to existing private institutions to establish "Agricultural & Mechanical" colleges, such as Texas A&M. These land-grant colleges are known then as "1862 institutions." Two additional Morrill Acts came along to support communities underserved by the 1862 institutions. The Morrill Act of 1890 established what are today known as "historically black land-grant universities." Most of these came about in the Civil War aftermath in the South and include institutions such as Alabama A&M University and Howard University in Washington, DC. The Morrill Act of 1994 set aside monies to establish institutions for Native American populations.

This educational goal is what funds and fuels our university Extension system today, providing farming and other related education at the county level.

Extension is here to serve you for free — your tax dollars at work. Extension hosts a wealth of farming resources and education opportunities from enhancing soil health to home pickle processing. Extension "agents" are university employees who focus on different areas of expertise. An agriculture agent focuses on farming-related topics. Extension may also have a horticulturist or community development person on staff.

As part of the nationwide Cooperative Extension System, your state's Extension is able to draw on research-based expertise from land-grant universities across the country. The challenge can be that if you are in a conventional agriculture area, like me, your agriculture agent may know only about large-scale, chemical agriculture practices. Terms like "organic" may be out of his box.

Yes, I use the term "his" intentionally, as most agricultural agriculture agents are men. That said, as the sustainable agriculture movement grows and more women are at the forefront, things are changing at Extension. Stay tuned. More outreach and programming targeting farmers like you will continue to grow. Feel free to jump state borders and check out Extension information and farming resources from other states. Some great options for small-scale farming information are in Vermont, Pennsylvania, California, and Oregon.

3. Grassroots nonprofits

With the growing organic market and farmer movement, almost each state or region now boasts a nonprofit specifically dedicated to that agricultural area. Some of these nonprofits are state specific, such as the Pennsylvania Association for Sustainable Agriculture (PASA); others focus more regionally, such as the Northeast Organic Farming Association (NOFA). Most have an organic specialist on staff or can connect you to a resource.

The National Sustainable Agriculture Coalition (NSAC) represents these sustainable agriculture organizations collectively on Capitol Hill and amplifies our voice when it comes to lobbying for federal funding priorities. Check out their member roster for a snapshot of leading nonprofits around the country.

4. Books, blogs, webinars, and online resources

With an Internet connection and library card comes an easy on-ramp to access beginning farmer education. From first-person beginning-farmer blogs sharing information to encyclopedic tomes of how-to information, pick your portal for free or low-cost learning.

"Remember, in navigating the resources out there, to read with a discerning eye," advises Harriet Behar of MOSES. "Gather multiple perspectives and evaluate your source. Read a beginning farmer's blog post on their experiences dealing with invasive species or drought and then see what NRCS or an organization like the Rodale Institute has to say."

5. Hands-on learning: internships and apprenticeships

You can read all the how-to-raise-chickens books out there, but nothing beats the first time a chicken poops on your food. Test the waters with a variety of on-farm internships and apprenticeships, which vary in length. The World Wide Opportunities on Organic Farms (WWOOF; wwoof.net) lists internships on organic farms around the world. ATTRA also houses an extensive internship listing (attra.ncat.org).

Access land: finding your on-ramp

"The most important thing to remember when thinking about farmland is what your goals are for the property," shares Ann Larkin Hansen, author of *Finding Good Farmland: How to Evaluate and Acquire Land for Raising Crops and Animals.* "This will direct everything from how many acres to the type of terrain and soil you'll need to location." Location can play an increasingly

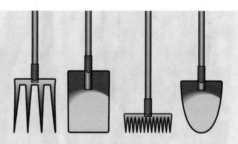

Tool Shed:

Female Farmer Words of Wisdom

We farmers love books. While *Soil Sisters* is the first book dedicated to farm training and start-up through a gender lens of women committed to organic sustainable agriculture, several female farmers have penned fabulous beginning-farmer resource books that you can share with the men in your life:

The Organic Farming Manual: A Comprehensive Guide to Starting and Running a Certified Organic Farm, by Ann Larkin Hansen

Farms with a Future: Creating and Growing a Sustainable Farm Business, by Rebecca Thistlethwaite

Sharing the Harvest: A Citizen's Guide to Community Supported Agriculture, by Elizabeth Henderson and Robyn Van En

Woman-Powered Farm: Manual for a Self-Sufficient Lifestyle from Homestead to Field, by Audrey Levatino

Market Farming Success: The Business of Growing and Selling Local Food, by Lynn Byczynski

Sustainable Market Farming: Intensive Vegetable Production on a Few Acres, by Pam Dawling

The Mobile Poultry Slaughter House and *The Food Activist Handbook,* by Ali Berlow

expensive role as we continue to lose one acre of farmland every minute, and farmland prices doubled between 2000 and 2010, according to American Farmland Trust.

Hansen recommends starting out your farmland access journey by asking yourself four questions:

• Do you want to generate more of your income on or off farm?
"Ask yourself if access to the job or good farm infrastructure and markets have the priority."

• What would you like to produce?
"If your goal is to sell farm products, you'll need land appropriate for your enterprise as well as good markets."

• What are the other people involved in your farming venture looking for? What's not a priority to you might be to someone else in your family. You may like the isolation of rural living, but that adds to a longer commute for your spouse. Children, whether you have them now or might in the future, may have a long bus ride to school. We'll explore more on integrating family and exploring options like homeschooling in Chapter 9.

• How will these needs and wants change in the long term?
"Life changes and, knowing no piece of property is perfect, how can your land best adapt and evolve with your future dreams and goals?" sums up Hansen. You may not be looking at livestock from the start, but if grass-fed beef intrigues you for the future, make sure you have adequate acreage for pasture.

What Farms Need

Different products require different soils, terrain, infrastructure, and markets. If you know already what you want to produce, use this chart to see what your farm will need.

Product	Acreage	Terrain	Soil	Water Needs
Vegetables, small fruits	0.5 acre (0.2 hectare) minimum	Level	Fertile, deep, well-drained	Abundant and reliable for irrigation system
Orchard fruits	5{+} acres (2{+} hectares)	Level to gently rolling	Moderaterly fertile, well-drained	Reliable, good quality
Field crops	Moderate to major	Level to gently rolling	Reasonably fertile, well-drained	Varies by crop
Dairy	Moderate to major; must have pasture (to satisfy organic requirements)	Level to hilly if well managed	Adequate for pasture	Abundant and reliable
Poultry	Minimal	Level to rolling	Good for improving soil	Adequate
Pigs	Minimal	Level to rolling	Good for improving soil	Adequate
Sheep, goats, cattle	Moderate to major; must have pasture	Level to hilly (if well managed)	Good for improving soil	Adequate

Excerpted from Finding Good Farm Land ©
ANN LARKIN HANSEN

So you've found the ideal property of 30 acres, which happen to butt up against federal forested land that has much lower odds of development. It has a charming century-old farmstead that needs some upgrades but isn't a money pit. There's also an older dairy barn and shed in decent shape along with a newer machine shed. Game on.

"The two most common ways to access land include owning it or leasing it," explains Terra Brockman of The Land Connection, a central Illinois nonprofit providing farmer training and public education on local, sustainably raised foods, and author of *The Seasons on Henry's Farm*. Ownership could come through family inheritance or an outright purchase.

"Ownership is appealing as it has psychological benefits along with security, control, and collateral building, but this route may tie up capital you might need for farming and requires a bigger commitment to stay put," offers Brockman.

With a lease arrangement, you pay rent for the use of the land and potentially other farm infrastructure. "There are many kinds of leases, but all good leases have equitable and clear division of rights and responsibilities between owner and farmer, plus a trigger for a lease review and exit provisions." Leasing from a land trust or from a farmer who shares your sustainability priorities often serves as a real plus. Leasing gives you a lot of flexibility in terms of not needing to put as much money into the operation up front and enables you to try out a location or enterprise before committing.

There are also lease-to-own options, including a shared-equity purchase, which involves allowing for a gradual transfer of your money and the owner's farmland, with small regular cash outlays made by you, the renter, and gradual capital gains for the owner. As your new farm gets more profitable, you can buy more land.

Incubator farms

A kind of lease arrangement with benefits, incubator farms give you a business start-up boost without the infrastructure cost. Typically run by a nonprofit, incubator farms provide land access to multiple farmers on the same property as well as consulting, training, and other resource support — often for a nominal rental fee.

The National Incubator Farm Training Initiative, out of the New Entry Sustainable Farming project at Tufts University in Massachusetts, supports farm incubators, now over one hundred across the country. More than half serve immigrants and refugees, but others serve a range of new farmers, from college-educated youth to career-changers or seniors looking for retirement income.

Cash crops: loans, grants, and crowdfunding

Loans

Unless you have a sizable nest egg or inheritance, you probably immediately think of option one: Go to the bank and get a loan. If you are still employed with a regular income and can show W-2s, have a good credit report, and can make a decent down payment, you could potentially receive a standard mortgage that would cover you, depending on your property specifics. That's what we did back in 1996 when we purchased our place. Little did the bank know we were quitting those jobs and moving to the farm full-time. We never missed a payment, and we ended up paying off our mortgage more than ten years ahead of schedule, saving thousands of dollars of interest.

Today's post-recession banking world, with changes to Fannie Mae, makes that process harder if you don't fit into simple application check boxes of being gainfully employed. If you're self-employed already, young, or coming into the farming world without assets, you may encounter barriers.

If you don't qualify for traditional bank loans, knock on the FSA's door. The FSA likes to call itself the "lender of first opportunity," but it more likely is "the lender of last resort." This means these loan programs are intended for farmers who are unable to obtain a loan through a traditional bank because of things like poor credit history or lack of the required down payment. If a bank will give you the full loan to purchase and start your farm operation, you probably won't qualify for an FSA loan. For this reason, these loans can be extremely appealing to young women who have experience, enthusiasm, and a plan but lack financial capital.

Once qualified through FSA, women farmers receive priority status among available funding because we qualify as an "SDA" (socially disadvantaged) farmer group. SDA farmers also include groups such as African Americans, Hispanics, Native Americans, and others who, for various reasons, have been discriminated against in the past.

Another big appeal of the FSA loans is a low interest rate and more appealing terms than at a regular bank. The FSA also has flexibility to work with young farmers who do not have savings toward a down payment. And the FSA understands and supports farm businesses from a loan payment perspective. For example, there is flexibility on payment schedule, and payments can be based on the type of enterprise and when produce or livestock is sold, rather than regular monthly amounts.

An underlying requirement for FSA federal farm loans is that applicants plan to farm as their primary source of income and have the business plan to support it. For example, rural residences operating small, part-time businesses would not qualify for these loans.

To apply for an FSA loan you must have a minimum of three years' experience in the agricultural area you are going into. The FSA has broadened its definition of "experience" beyond a degree at an agriculture college to include the "portfolio" idea of amassing various experiences including programs like Farm Beginnings and Annie's Project, online curriculum, internship and apprentice programs (particularly with management responsibilities), the MOSES Farmer-to-Farmer Mentoring program, and events like EcoFarm and the National Women in Sustainable Agriculture Conference.

Keep records and track your experiences. Get reference letters from internships. If you ever apply for a FSA loan, you will need such documentation, and it's much easier to simply ask for that while you are there.

Allocate time for the approval process. Unlike at a bank, the FSA loan approval process can take much longer, which might adversely affect a relationship with a potential seller. For example, the FSA can start processing a loan application only once you have an accepted offer, which can be less appealing to the seller who wants to close the deal. So it's important to start the process early, develop a relationship with your FSA loan officer, and have your paperwork ready to go when the offer is accepted.

If you think an FSA loan might be in your future, the first step would be to research the various programs, both for farm purchase and capital acquisition, available within the FSA. "Your Guide to FSA Farm Loans" is an easy-to-read, online resource that

Idea Seed:

"Check whether local cooperatives such as your food co-op might have loan programs. A group of local farmers and I together applied for and received one for three thousand dollars from our Chequamegon Food Co-op to purchase a used walk-behind tractor. As the co-op supports area farmers, the loan was at a very low interest rate and on very favorable terms for repayment over two years. It's a piece of equipment we can readily share as we all need it intermittently, and it's small enough that we can trailer it between farms or if it needs to go to the shop."

— Clare Hintz, Elsewhere Farm, Herbster, Wisconsin

synthesizes loan and loan serving options (www.fsa.usda.gov/dafl). If you generally qualify and think this may be in the future for you, the next step would be to contact the FSA loan officer covering the area where you want to purchase property.

Grants

Various funding entities can potentially add dollars to fund the farm. Unfortunately, I have yet to see any grants that directly fund farm purchases or that are specifically for women farmers. However, grants do exist that will help fund a specific component to expand your business once you have your farm up and running.

"The most important component to a successful grant application is the same as a solid farm business plan. You need to research and write out a strategic blueprint for your farm vision and the steps to successfully get there," explains Traci Bruckner with the Center for Rural Affairs in Nebraska, a nonprofit organization advocating for funding for programs supporting sustainable agriculture. "Federal grants to support high-value, niche markets are highly competitive, so as an applicant, you need to be thinking big picture as well as finite details, answering the tactical questions that need to be addressed, and thinking through where you see your farm business heading."

What makes a successful grant proposal? Here are three key steps:

1. Strategize before writing

"First, make sure your project has a tightly defined purpose along with a clear strategy to accomplish your goal and a defined and realistic timeline," offers Margaret Krome, policy program director at the Michael Fields Agricultural Institute in Wisconsin. Krome conducts workshops around the country, helping farmers access federal programs that support sustainable agriculture. "Make sure your plan includes the people, money and other resources needed to accomplish your intent, a basis for evaluating the process when done and an effective means of communicating results to any audience that needs to hear them."

For Beth Osmund of Cedar Valley Sustainable Farm, grant funding enabled her and her husband, Jody, to strategically research, test, and prosper by becoming the first meat CSA servicing Chicago. They received a "farmer rancher grant" through their North Central Region of Sustainable Agriculture Research and Education (SARE), a federal program supporting various aspects of sustainable agriculture. These particular competitive grants fund farmer and rancher-led projects that explore sustainable solutions to problems through on-farm research, demonstration, and education efforts.

"We had been doing a diversified vegetable CSA since we started in 2003, with some livestock on the side," explains Osmund. "We felt that the meat side of the business could be more lucrative and a better fit for our farming interests and lifestyle. Our SARE grant helped us explore how to best market our meat using a direct-to-consumer model through venues like farmers' markets." Their SARE grant included compensation for the Osmunds' time to do things like develop a meat product line and research pricing.

"This grant enabled us to try out and test new marketing venues we wouldn't have been able to do otherwise," adds Osmund. Exploring the idea of providing meat at farmers' markets, the duo came up with the idea of using the CSA model for meat, selling monthly "shares" of beef, pork, and poultry cuts.

2. Choose suitable programs

Krome advises asking the following questions when evaluating a grant opportunity:

• What is a program's stated mission and objectives?
• What is a program's funding pool, percentage of applicants who typically get funded, and average amount and duration of grants?
• What are the eligibility requirements, financial match requirements, and restrictions on a program's use?
• Are a program's application deadlines and funding time frames suited to your timing?
• What do past grantees say about how the program is administered?

After you identify your farm funding needs, start researching potential grant options. The free download "Building Sustainable Farms, Ranches and Communities," developed by the Michael Fields Agricultural Institute, provides a concise overview of various federal funding sources. The National Sustainable Agriculture Coalition also offers a free download, the "Grassroots Guide to the Farm Bill," with a synopsis of Farm Bill programs related to funding options and also offers a free weekly email with farm program updates.

Typically when a particular grant is open for applications, the administering entity will issue what's called a "Request for Proposals" (RFP) or "Notice of Funding Availability" (NOFA). This information, available on the organization's website, will outline the details, including how the proposal needs to be written, deadlines, and what the grant will and won't cover. Some grants are more educational in focus and will directly fund only nonprofit organizations. Others, like SARE, will make a grant directly payable to the farmer.

"Read the RFP several times to be sure you understand exactly what the particular grant will and won't fund," advises Krome. Be particularly keen on understanding what types of applicants qualify for that particular grant; are farms eligible? Is it suitable for a farm of your size and nature?

Think beyond the expected agriculture grant resources and see if there are ways to "connect the funding dots" with other programs, particularly opportunities in your state and local area. Holly Mawby, co-owner of gardendwellers Farm, a culinary herb and agritourism operation in North Dakota, did that through tapping into various local groups, from utilities to tourism and historical societies. "We received funding from our state historical society toward an outdoor interpretive sign for our farm," Mawby explains. "Our farm now helps support telling the story of our area's agricultural heritage. We now offer the historical society another spot to add to their tour map."

"I'm always connecting the dots," Mawby shares with a smile. "I'm always reading local papers, and whenever I see an announcement of a grant being awarded, I think about how that might fit our farm." Mawby recommends tapping into organizations where you are a member and use their services, like rural telephone and electric cooperatives. "Help these groups see how your farm shares and supports their mission, like preserving area history or spurring the local economy with new jobs or tourism."

3. Write a straightforward proposal

Once you decide to proceed, read the RFP at least three times. "Pay close attention to format, deadlines, and requirements," adds Krome. "Don't exaggerate the need or over-promise results, and have your proposal reviewed by someone whose editing and thinking skills you trust."

Take time to research and write a detailed and appropriate budget. Does the grant require "matching funds?" Matching funds are financial resources you need to bring to the plate to "match" the funds you ask for, as an indicator of your commitment to the project. Ask questions about exactly what qualifies as a "match" since "in-kind" contributions, such as existing equipment or volunteer labor, can often be included as your matching funds instead of cash.

Allocate more time than you think you need to write the grant. Sometimes grant applications require additional elements that take time to gather, such as letters of support or tax records. Keep abreast of deadlines. Don't assume that because a grant targets farmers it won't have a deadline in the peak of summer. Unfortunately, government legislators don't think like farmers. Many grants have the same annual general timing cycle, so you could use downtime in winter to write the bulk of an application that may be due during peak planting months. That said, USDA programs in particular are tied to the federal budget and can vary in funding allotments year to year. Don't assume the parameters this year will be same as next.

Some grants may require you to involve partners and other community organizations directly in the project and application process. For example, Mawby has found that when certain grants require the applicant to be a non-profit organization, which her farm is not, she can partner with a local group that has 501(c)3 status to officially submit and be the lead applicant. Her farm takes on a partnering role. She still receives the financial backing; however, it is paid through a partnering organization.

Don't wait till the last minute to submit that application. Lots of issues can crop up that could jeopardize your whole application, especially if they happen at the last minute. Build in some easy "insurance" by submitting the application a few weeks, or at least a few days, before the deadline. Realize, too, that with the increasing number of grants requiring online applications, there may be lots of people submitting proposals at the last minute, which could create technical hiccups on the government servers and cause your proposal to miss the deadline.

Approach grants like you plant the fields. Always plant a few extra seeds because you know not everything will germinate. It's the same thing with grants. Resources are tight and most grants are competitive. Rejection is a reality of the process but one to learn from.

"If you don't receive something you applied for, by all means ask questions and garner an understanding of why things didn't work out. Many successfully funded proposals are the result of past rejections," advises Krome. Call the staff and politely ask questions. It may be that something was off-target or missing from your application. Or it may simply be that you didn't have the strongest proposal and the process was very competitive. Don't argue with the grant administrator or express severe disappointment. Consider post-rejection discussions as simply an open door to gather helpful feedback for next time.

If you do receive the grant, after you pat yourself on the back for an application well done, make sure you fully understand next steps and any additional requirements you need to complete. Some grants necessitate you to sign a contract and complete ongoing specific reporting forms for several years. Take the time to process these details before jumping into the project to avoid hang-ups — and possible payment issues — down the line.

"We need farmers to both use these programs and speak up when federal funding is on the line," challenges Krome. "It's also important for farmers who receive support to share their learning and success to help further champion the sustainable agriculture movement."

Idea Seed:

"I applied for and received a Value Added Producer Grant (VAPG) to take my herbal business to the next level. I was producing dried herbs for teas in bulk, and this grant enabled me to start producing tea in bags, which tremendously increased my wholesale market as most Americans only buy tea in bags. I'll be first in line to say this was not 'free money,' as not only did the application take work, but once we received it everything amped up. That said, this grant catapulted our farm in a way professionally that we would never have been able to do on our own."

— Jane Hawley Stevens, Four Elements Organic Herbal Farm,
North Freedom, Wisconsin

Crowdfunding

You've probably heard of crowdfunding platforms like Kickstarter or Indiegogo. These are portals that enable you to draw funding support from those who believe in you most — your customers, friends, and family. You have an opportunity to broadcast your message to the general public so that strangers are lured to become supporters, too.

Crowdfunding generally consists of setting a fundraising goal and then offering various incentives and prizes for different levels of payment. Supporters pledge online and then payment is made if and when your fundraising goal is reached. Crowdfunding can be an amazing way to tell your story and get support from your community and the greater world beyond. It's all part of the growing "sharing economy" movement that we all need to be connected economically directly to each other.

"Crowdfunding can work exceptionally well if you have an existing customer support base familiar with your business, like a devoted CSA membership," comments Jan Joannides, executive director of Renewing the Countryside, offering resources that champion new financing models for farms. "If you don't have that existing backing, you'll need to focus more on creative outreach to grow your message further."

"Women farmers are a natural for crowdfunding as the process draws on a lot of the strengths we naturally have," shares Maria Sayles, director of projects and community at Barnraiser, a crowdfunding platform specifically focused on farm and food ventures. "We're good at nurturing community and connecting authentically, which form the core of Barnraiser: championing support by honestly sharing our vision and mission."

Sayles sees a common crowdfunding pitfall as setting the fundraising bar too high. "Don't ask for fifty thousand dollars if your capacity is not there yet." For a successful campaign on Barnraiser, typically one third of funding comes from family, friends, and direct network. The next third comes from "friends of friends," people one or two degrees away who hear about you through your direct network. The last third is from the Barnraiser audience and the people who find you via the Internet, social media, or other venues. "Don't forget, too, that there are benefits to crowdfunding beyond the actual cash ... as you will build new audiences, make new partnerships, and gain media exposure through your campaign," adds Sayles.

Here are some tips from Sayles on how to succeed in crowdfunding platforms like Barnraiser:

1. **Find your crowd and audience**

   For crowdfunding, your audience is made up of all the people you can reach to tell about your project, including family, friends, customers, fans, clients, online groups, business contacts, community groups, membership organizations, and anything else. Email is the best way to reach people, so start collecting email addresses and building alliances before the start of your campaign.

2. **Tell a compelling story**

   Does your story have broader impact? Is it a model that others could replicate? Include these elements in your project page and try to make the story relevant for folks outside your immediate community.

3. **Set a realistic goal**

   Sure, everyone could use $100,000, but remember that most projects will raise between $5,000 and $25,000. Only about five to ten percent of the individuals on your email list will donate, and a typical pledge is between $50 and $100.

4. **Give great rewards**

   Believe it or not, the rewards are what draw a lot of backers in. Give compelling rewards set at a reasonable price and remember to consider the cost of time and shipping.

5. **Use a video**

   Post a one- to two-minute video, especially if you are trying to raise over five thousand dollars. Professionalism in the video carries over to respect for the campaign, but this doesn't mean it has to be a blockbuster hit. Instead go for short and compelling, with good shots and clear audio.

6. **Be fearless with the ask**

   The project creator and team are their own biggest cheerleaders and cannot be afraid to get the word out. It's only for a month or so, and it's for a great cause — you — so everyone needs to be ready to promote and push.

By now you're ready to start planting, to start moving ahead with the actual vision for your farm livelihood. In the next section, we'll do exactly that as we cover developing your business plan.

PART 3:
PLOWING AHEAD

Supplying the bank account and sustaining the soul

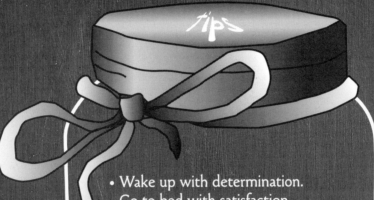

- Wake up with determination. Go to bed with satisfaction.

- Explore and experiment.

- Remember that there is always an elegant solution to every problem.

- End your days with an Epsom salt bath.

- Carry a notebook everywhere.

- Bless your harvest.

- Trees can grow through rock.

- Have a friend take quality, high-resolution photos of you at work.

- Print professional business cards with your name and title: farmer.

Chapter 5

Farm Business Varietals

O H KITCHENAID, how do I love thee? Let me count the ways. You stir, mix, beat, cream, and whip egg whites to peak perfection. Add in a couple of accessories and I can stuff sausage, grind grains, crank out ice cream, and pump out pasta that even Giada De Laurentiis would love. But more than a trusty workhorse, you are so stylish that the San Francisco Museum of Modern Art keeps you on display as an icon of American design.

Perhaps you share my KitchenAid love? I have a soft spot for reliable multi-function items that look good. Heck I'd put my husband, John, in that same category. He never falls short on a promise, and can both rototill and roll out a Martha Stewart-worthy piecrust. As he nears the half-century mark he's falling into that George Clooney-esque category of men who keep their hair and age well.

From KitchenAids to walk-behind tractors, versatility makes a situation stronger. By keeping farm operations diverse, with multiple income streams synergistically flowing, we grow stronger enterprises that can survive when the blight hits the tomatoes. Look a few rows over, and the basil is having the most prolific year ever, fetching a premium price at market. The Scottish Highland cattle in the field support pasture health with their manure.

This chapter focuses on the core of any farm business: agricultural production.

Consider this the KitchenAid base unit as you plow ahead planting business seeds. In the next chapter, we'll cover diversification further through complementary on-farm businesses such as farm stays and freelance writing. Like the

Lisa Kivirist with her beloved KitchenAid.

JOHN D. IVANKO

juicer and sausage stuffer accessories for the KitchenAid, these entrepreneurial additions add income through diversifying the core base of whatever it is you produce on your farm.

Grow it: field production and tree fruits

"Keep it simple and set yourself up to succeed when you get started because there's nothing more empowering than a flourishing first couple of harvests," advises Pam Dawling, author of *Sustainable Market Farming: Intensive Vegetable Production on a Few Acres.* An avid vegetable grower, she has been farming as a member of Twin Oaks Community in central Virginia since 1991, where she helps grow food for about 100 people on three and a half acres. Born in London, Dawling became part of the back-to-the-land

Pam Dawling in the hoophouse.

DENNY RAY McELYEA

cooperatives in the early 1970s and is among the early visionaries of cooperative organic farming as a healthy way to live and an alternative to industrialized agriculture. "Growing food should never feel isolating or overwhelming if you approach it as something done with a collaborative, joyful spirit and in partnership with both Mother Nature and community."

From a tiny seed a mighty livelihood can grow in crops that the USDA defines as "specialty crops." Dawling offers five seeds of wisdom to get you started:

1. Launch with what's easier

"Start with the vegetables that are easiest to grow in abundance, like greens," Dawling recommends. Rich in antioxidants, fiber, and many other essential nutrients, leafy greens, like kale, spinach, Swiss chard, and collards can be planted for both spring and fall crops and, depending on your climate, can often be harvested throughout the summer too.

"Focus on growing those greens that you like to eat as it gives you a lot more positive energy in the growing fields and you always have a market for

Greens in the garden.
JOHN D. IVANKO

excess harvest: yourself." Dawling also advocates planting some easy herbs like chives and cilantro.

2. Go public at farmers' markets

"Think about starting out those first couple of seasons selling at a farmers' market because there's no pressure to deliver certain specific harvests. You can just bring whatever it is you have," Dawling advises. Don't jump into a CSA right away as the pressure increases at the get go to distribute weekly expected harvests to paying customers. Get your business feet wet at the farmers' market for a few seasons to determine what it is you both like to raise and can grow well. "A CSA may very well be in your farm's future, but the pressure and stress lessens when you can confidently start a season knowing how much you can deliver when and build your boxes around that."

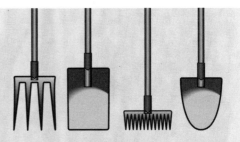

Tool Shed:

Size by Definition

For small-scale farm operations, the following definitions are commonly used to put a label on your enterprise:

Market gardens:
Fewer than three acres in active production, not including fallow or cover cropped areas. Most work is done by hand. Market garden farmers produce a diversity of vegetable, fruit, and/or flower crops to sell directly to consumers.

Market farms:
Between three and twelve acres in active production, not including fallow or cover cropped areas.

At this size, issues of mechanization versus hand labor and what makes the most cost-efficient sense can challenge farms.

Vegetable farms:
Produce crops on more than 12 acres, not including fallow or cover cropped areas.

Source: Center for Integrated Agricultural Systems, University of Wisconsin-Madison College of Agricultural and Life Sciences

3. Love those perennials

"Plant once and harvest for years. What's not to love about perennials?" offers Dawling with a smile. Asparagus, artichokes, herbs, and berries hold particular promise because of their relatively short establishment time, with some varietals harvestable in their first year. "These fruits and vegetables are also popular among market customers and can bring in a higher market value."

4. Balance your growing calendar

"You can control various elements of your growing schedule and think about planning open blocks when you don't have obligations in the field and can take some time off," Dawling suggests. "The soil needs a rest and so do you. So plan your growing timeline so there are some blanks where you can get away."

For my Inn Serendipity operation, we focus on an intense growing season from spring through late summer. By August, most of our crops are in except for winter squash, leeks, and some fall varietals. Sure, I could extend the season into autumn and even later with hoop houses and cold frames, but I like a shorter, concentrated growing focus. By early September I'm physically and mentally done for the season. As we're just growing for our B&B guests and some local outlets and our own needs, we can raise enough under this schedule, go full throttle on food preservation

Idea Seed:

"During the harvest season, I'm high on produce abundance and low on time to do any home preservation for my own family's needs. One of our regular farmers' market customers loves to can, so he takes whatever he wants from our remaining inventory at the end of market, cans it and shares half of the jars with us. It's an easy win-win for everyone and I can slowly savor my own bounty all winter long."

— Peg Sheaffer, Sandhill Family Farms, Grayslake, Illinois, and Brodhead, Wisconsin

Idea Seed:

"Raising flowers particularly appeals to women as we deeply connect with and appreciate the beauty of these blooms. Flower growing can also be exceptionally compatible and complementary to family life with children, as you harvest in the cool of the day — very early in the morning or late in the day — times when young kids are often asleep."

— Lynn Byczynski, editor of *Growing for Market* and author of *The Flower Farmer: An Organic Grower's Guide to Raising and Selling Cut Flowers* and *Fresh from the Field Wedding Flowers: An Illustrated Guide to Using Local and Sustainable Flowers for Your Wedding*, Lawrence, Kansas

Andrea Clemens with her flower harvest.

JOHN D. IVANKO

and then step aside from major field duty into the late summer. We'll even get off farm for some summer travel in late August, which would be impossible if we committed to more fall crops or a CSA.

5. Keep notes and document your journey

"Your best teacher is your own past experience, so make sure to take notes and keep records of everything you do," Dawling sums up. She keeps detailed crop records and notes on better methods, varieties and timing. Over the decades, this has added up to a treasure chest of practical knowledge. "But don't keep your insights to yourself. Share your successes and tribulations with others." Dawling does exactly that as she summarizes her insights into monthly articles for *Growing for Market* magazine, reflections that eventually became the foundation for her book, *Sustainable Market Farming*.

Another approach to field crops: Focus on the unusual and different — items that no one else is selling. For example, luffas. "Everyone always asks, 'What is that?' when they first see these gourds," explains Deanne Coon of The Luffa Farm with a grin. "But luffas used as sponges come from a long history, reaching back to early Egyptian days." They are a fibrous, sponge-like gourd that matures into a hard mass of fiber that can be used as a scrubbing sponge.

"The luffa is an amazing visualeducational tool as it completes its whole life cycle on the vine and is quite the intelligent plant," Coon continues.

Starting as an enthusiastic hobby over 25 years ago, her luffa passion led her to open her farm in 2000 in Nipoma, California. In addition to offering complimentary tours, she sells various luffas, luffa seeds, and a range of natural soaps and lotions she handcrafts on site. Bonus: As luffas are not consumed as food, her products do not need to be made in a commercial kitchen; nor are they subject to state regulations.

"Think of luffas as a cross between snowflakes and orchids: No two are alike," Coon shares. "They are addictive and fun to grow." A veritable Johnny Appleseed of luffas, Coon spreads her luffa love with seeds and advice to visitors to her farm, using the same heirloom seeds she started with decades ago.

Raise it: livestock

Animals do wonders for small-scale farms on multiple levels. Their manure adds nutritional gold to soil quality; they provide meat, eggs, or milk for both home use and sale; and — let's face it — they can be darn cute and entertaining. But animals of any sort catapult you into a deeper level of farm commitment, with daily chores and responsibilities. It's one thing to take a spontaneous trip off the farm and come home to extra weeding in the field. If a couple of plants die, so be it. With animals,

Deanne Coon of The Luffa Farm in Nipomo, California. JOHN D. IVANKO

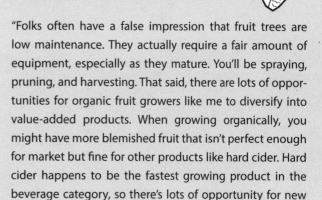

Idea Seed:

"Folks often have a false impression that fruit trees are low maintenance. They actually require a fair amount of equipment, especially as they mature. You'll be spraying, pruning, and harvesting. That said, there are lots of opportunities for organic fruit growers like me to diversify into value-added products. When growing organically, you might have more blemished fruit that isn't perfect enough for market but fine for other products like hard cider. Hard cider happens to be the fastest growing product in the beverage category, so there's lots of opportunity for new business start-ups. Cheers!"

— Deirdre Birmingham, The Cider Farm, Mineral Point, Wisconsin

How She Sows it:

Cathy Linn-Thortenson, Wise Acres Farm, Indian Trail, North Carolina

Very Berry Wise:
Starting with Strawberries and Community

"I have to laugh whenever I fill in the checkbox next to 'farmer' when it asks me my occupation," shares Cathy Linn-Thortenson with a grin. "I still can't quite believe the changes that happened in last couple of years and that we're actually here on the farm."

"Here" for Linn-Thortenson and her husband, Robb, and their three kids is 40 acres just outside Charlotte, North Carolina. Officially opened to the public in 2014, Wise Acres Farm is a certified organic U-pick strawberry operation. Linn-Thortenson's journey to farm life started 800 miles northwest in a place far removed from the soil: a corporate cubicle in Chicago.

Cathy Linn-Thortenson of Wise Acres Farm.

JOHN D. IVANKO

"We were complete urban dwellers caught up in the business career scene," explains Linn-Thortenson, who earned her MBA at the University of Chicago while Robb worked as a commodities trader. "I loved the challenges of the business scene but came home feeling disconnected and, bottom line, not feeling good about what I accomplished that day."

After a few years of enthusiastic urban gardening ventures growing in recycled pickle barrels and, as Linn-Thortenson puts it, "watching one too many depressing food documentaries," the wheels started turning toward a new life chapter in farming. "At first, we figured we'd get a small hobby farm but then realized, why just keep this a hobby? We totally believed in growing healthy food and the importance of connecting people,

especially families, to their food sources. Let's jump in fully and make this our family business venture."

High land prices in the Chicago area led Linn-Thortenson south, eventually to North Carolina. While they were looking at land options, she and Robb embraced researching the venture with gusto. "That's where our business background helped. We knew we needed to be strategic in what we raised, especially given our limited growing experience. We also needed a readily accessible market for our crops."

Those two variables led them to consider U-pick strawberries. Strawberries made sense since they could be planted in October. They'd be ready for picking the following April. Turns out, there was only one other organic strawberry grower in the state. "U-pick strawberries also gave us the perfect outlet for fulfilling our mission to connect kids to their food sources and learn about caring for the Earth. There's nothing like picking your own sweet berry for some real-life learning!"

Wise Acres Farm.
JOHN D. IVANKO

Linn-Thortenson needed to be open to move quickly when opportunity arose, or as in this case, when the snow came. "There was a farm auction near us that we were planning to go to. It snowed that day, which keeps people home in North Carolina. Robb went and, because there weren't many attendees, kept buying up equipment we needed at incredible prices."

This vision for a U-pick strawberry farm and the desire for their kids to connect with the types of schools and activities they were used to in Chicago led Linn-Thortenson to a 40-acre parcel on the outskirts of Charlotte, just beyond the suburbs but accessible for families looking for an afternoon on the farm. "Our kids, all under twelve when we moved, were quite active in sports, and we wanted to make sure they could continue that in our new place." The whole urban-to-farm transition,

from the research phase to moving in, took about two and a half years. "It took us those last six months when we were already on the farm just to clean up the place. It was in pretty sad shape and had been in foreclosure," admits Linn-Thortenson.

Linn-Thortenson currently has about one and a half acres in strawberries, which adds up to an intensely busy few months from April through early June. "I feel like the bad mom when the kids' soccer season starts as I never make a game in May, but after the strawberry season wraps up, we have the whole summer to connect as a family."

Linn-Thortenson and Robb choose strawberry varietals based on when they ripen so they can expand the U-pick season as long as possible. "Our challenge has been educating customers on the strawberry growing season and that it's relatively early in the spring. Folks still come in July looking for berries."

Social media proved to be the key marketing outlet for Wise Acres. "I was never much of a Facebook person personally, but it totally made our first season work since that's how folks heard about us." Linn-Thortenson's healthy food priorities extend to the snacks they sell on the farm. "I'm all about fun treats for the kids, like healthy, organic snacks of popcorn, lemonade, a strawberry slushy without corn syrup, and cotton candy without the artificial colors. Kids are amazed that the cotton candy is white!"

It's hard to picture Linn-Thortenson, tall and willowy and rugged around the edges, in a corporate suit. She exudes warmth and welcome, a passion for the farm that comes across vividly to each customer driving up the gravel to her farm. With strawberries established, she's looking at further diversifying into crops with higher margins, such as blueberries and mushrooms. She wants to keep the income coming in year round to smooth out her cash flow.

"Hands down, we wouldn't be here without the amazing support we've received from both the farming community and supportive state agencies," sums up Linn-Thortenson. "Extension agents have been incredibly helpful to us. Even though by growing organically we were out of their usual education box, Extension was always willing to research what they didn't know, and we learned together." Linn-Thortenson is also grateful for how openly supportive other area farmers have been to her new operation. "Other farmers, even those running similar U-pick operations, have been nothing but gracious and helpful to us. Farmers realize we succeed together; if the tide rises all boats get lifted together."

however, spontaneity takes a permanent back seat, as you need always to find quality, reliable care before leaving. That's help you may need to pay for, lest you burn out the goodwill of neighbors.

But for many of us, when we watch a doe give birth to triplet goats in the middle of the night, any hurdle along the way disappears. For others, biting into our own juicy steak or helping our kids learn responsibility through animal care magnifies the rewards of raising animals.

"If you're just starting with animals, I recommend getting two feeder pigs," suggests Deborah Niemann, owner of Antiquity Oaks, a diversified farm operation in Cornell, Illinois. A seasoned homesteading expert, she's also the author of *Homegrown and Handmade, Ecothrifty*, and *Raising Goats Naturally*. "It's then a six-month commitment till they are ready to slaughter, which isn't a huge time frame to test the waters and see if pigs are for you. Six month also lets you try out different breeds to find out which temperament is best suited for you." Keep a sense of farm humor through it all: Niemann names her pigs after celebrity chefs such as Julia Child and Rachel Ray.

Pork remains the biggest meat seller and profit center for Antiquity Oaks. "I raise and sell about twelve American Guinea hogs annually. I like that breed because the pigs are relatively small in size with a hanging weight of about 100 pounds, which is much more manageable for me." Bigger isn't always better when it comes to meat production. Niemann sells one whole hog directly to each customer, mostly clientele who live in the city and want sustainably raised meat, but don't have the freezer space to commit to, say, a Tamworth hog, which will weigh in at about 200 pounds.

"I find selling the whole animal directly much simpler and more cost effective," shares Niemann. "I take the whole animal to the locker and my customer takes it from there and can order exactly what cuts they want. I just need twelve people who each want one pig. That's tremendously easier than finding probably

Deborah Niemann of Antiquity Oaks with her goat friends.

100 people to buy all the bacon individually, plus I'd have all the other pork cuts that I'd then need to store."

"Make sure you're getting a fair price for your meat comparable to other farms in your area," advises Niemann. "I learned this the hard way as I was charging four dollars a pound for my grass-fed lamb, which I thought was fair, but once I realized others were charging much higher prices, I immediately raised my prices to five dollars a pound, and my customers didn't blink an eye."

Niemann's other animal passion is goats. Again, size matters when selecting your breed. "I like Alpine and Saanen goats as they are strong milkers. Nigerians have a high butterfat milk that makes fabulous cheese. These breeds are on the smaller size, at around sixty to seventy pounds, which makes them easier for me to handle. My goats and I need to see eye to eye, and they need to know I'm in charge," says Niemann with a laugh.

From your farm business perspective, goats provide a readily accessible income source: goat's milk soap. To sell the milk or make cheese commercially, you would need a full-scale dairy operation, which would run upwards of $250,000, according to Niemann. Soap making doesn't have any commercial kitchen requirements, and you probably have most of the equipment on hand already to start on a small scale: bowls, pots, and spoons. "My two starter investments for soap-making were a ten-dollar stick blender and a twenty-dollar digital scale."

"You can still use your goat's milk for your own home cheesemaking," adds Niemann, who makes eighteen kinds of goat's milk cheese, including cheddar and Gouda. She finds that men tend to raise goats for meat while women lean toward dairy. "But we women get a win-win that way as you automatically get meat with a dairy goat because dairy goats have to get pregnant every year to make milk, and you have to have something to do with the extra male kids."

"Do buy your goats from a reputable source and someone who is willing to be your mentor," offers Niemann. "Buy from someone who takes their goats seriously, which means they should have written milk records for dairy goats and know their average daily weight for meat goats."

Idea Seed:

"The most empowering thing I've done: learn to back up a trailer. Another woman farmer insisted that I do this. I can't tell you how many times that has gained me the respect of men at the livestock auction, feed store, or butcher shop. If you can back up that trailer with ease and finesse, you are a woman to be taken seriously."

— Jane Jewett, Willow Sedge Farm, Palisade, Minnesota

Let's move on to poultry, which are technically domestic fowl, including chickens, turkeys, geese, and ducks, raised for the production of meat or eggs. And let's start at the end this time: poop.

"When building a coop for any kind of fowl, make sure it's easy to clean," starts off Victoria Redhed Miller, author of *Pure Poultry: Living Well with Heritage Chickens, Turkeys and Ducks.* She lives off the grid on a 40-acre farm in the foothills of Washington's Olympic Mountains. "This comes in particularly handy during the freezing temperatures of winter when the days are shorter and the birds spend more time in the coop and you know what that adds up to more of. Plus, the easier it is to clean, the more likely you'll do it regularly."

Ducks, in particular, drink a lot of water and their poop is very wet. A lot. But poop aside, Miller advocates adding ducks to your business plan because their eggs sell for a premium. "Most people allergic to chicken eggs can eat duck eggs, so they are quite in demand and can garner a premium price of fifty cents an egg and upward," she explains.

"Chefs, particularly pastry chefs, love duck eggs because the whites have higher viscosity than chicken eggs, which means they will whip up higher and create loftier and higher cakes." Miller recommends the Khaki Campell breed, dubbed the "egg laying champion of the duck world." Overall, ducks are more prolific egg layers than chickens and take less time off from laying during the annual molt.

On the chicken front, think about New Hampshires, related to the Rhode Island Reds. Because they are fast to mature, in just eighteen weeks, Miller recommends this breed as a strong dual-purpose bird with a mellow personality. "You don't want to worry about breeds that might be too aggressive around little kids," she adds.

"A mistake that both beginners and more experienced farmers make — I know I do — is wishful thinking when it comes to animals," shares

Idea Seed:

"Alpacas can be quite efficient to manage, even easier to keep than goats. At one point I took care of twenty-one alpacas in one hour a day, including feeding and cleaning. They poop in the same place, which makes clean-up super easy. As long as they have pasture, hay, and some other alpaca buddies to hang out with, they're content and don't challenge fences. They're social animals. Once when I was eating dinner, I noticed an alpaca did get out of the pasture. I finished eating before I went to move him back into the paddock because I knew he wouldn't go anywhere. He'd just sit on one side of the fence waiting to get back to his buddies."

— Dr. Gail Campbell, veterinarian and owner of Ameripaca Alpaca Breeding Company, Galesville, Maryland

Jane Jewett, a third-generation farmer on family land in Minnesota. Entering into her 20th year farming, she runs Willow Sedge Farm and sells meat at the Grand Rapids Farmers' Market. "As a beginner, it's easy to get idealistic about being completely organic or raising livestock with low inputs or only using homeopathic remedies, but sometimes that leads to people not providing adequate or the right feed for good growth, or making other mistakes on housing, fencing, veterinary care, and so forth.

"Basic biology and forces of nature will rise up and smack you if you don't take a hard-eyed and pragmatic view toward problems like skunks and raccoons," Jewett adds. "If you think that because you spent a lot of money on your heritage-breed turkey

Idea Seed:

"When you find yourself yelling 'You sheep are so stupid,' stop. It's not them. You are the one doing something wrong. You're not thinking about how sheep move or what they need. Instead, you're expecting them to read your mind. Hopefully your sheep will be too polite to point out that this is just stupid."

— Catherine Friend, author of *Hit By a Farm: How I Learned to Stop Worrying and Love the Barn* and *Sheepish: Two Women, Fifty Sheep, and Enough Wool to Save the Planet*

Tool Shed:

Think Heritage Breeds

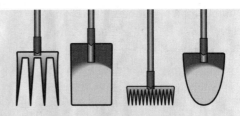

Ayshire cattle. Gloucestershire Old Spot hogs. San Clemente goats. Navajo-Churro sheep. As a small producer, you can focus on heritage breeds that folks won't find elsewhere. The Livestock Conservancy defines heritage breeds as:

Traditional livestock breeds that were raised by our forefathers [and foremothers]. These are the breeds of a bygone era, before industrial agriculture became a mainstream practice. These breeds were carefully selected and bred over time to

develop traits that made them well-adapted to the local environment and they thrived under farming practices and cultural conditions that are very different from those found in modern agriculture.

These unique breeds typically sell at a premium, as they are rare. Protect our agricultural heritage as well as your business bottom line.

For more on heritage breeds, see www.livestockconservancy.org.

poults the severe thunderstorm and hail will pass them by if you haven't built an adequate shelter for them, think again. If you're opposed to antibiotics and you think the cow with a severe case of foot rot will get better on her own because you're giving her minerals and essential oils, you may be disappointed." Jewett uses transparency as her farm marketing strategy: On her website, she lists every time she used antibiotics on one of her animals and why.

Create it: value-added products

Take a raw product raised on the farm and transform it into something else that you can sell at a higher price and — behold — you have a value-added

Cottage Food Pros and Cons

Pros	Cons
Little to no capital needed; you probably have everything you need in your kitchen already.	State regulations limit what products you can make, some more than others.
Fast start-up. Most states have a simple, low-cost registration process.	States may also have limitations on where you can sell; some do not allow special orders and restrict sales to farmers' markets or public events.
May already have a recipe and be experienced in what you want to make.	
Sell directly to the customer and keep more profit.	With any food product, you're liable for what you make and need to insure yourself for the risk you take.
There's nothing like the flexibility and freedom of being your own boss — you get to call your own shots.	Baking, canning, and other food preparation is hard work on your feet, especially if you have to make multiple fresh items at once.
You're helping build a stronger local economy and community connections.	Bookkeeping is a must since you're required to keep track of sales, expenses, and inventory. A real chore, if you don't like crunching numbers.
Defining success on your own terms.	
Opportunity to grow and expand *after* you prove a successful market.	May stir up some negative vibes when viewed as competition by local businesses like an established commercial bakery.

product. A variety of ways exist for you to create something of higher value out of existing raw ingredients, products as diverse as jams, salsa, cheese, dried herbs, and even painted gourds.

If your item is not intended for human consumption, such as soap or flower arrangements, you are not subject to regulations. Once we start talking food products, the rules, regulations, and expenses grow. A big factor is the type of product you're producing: The more complex, such as something refrigerated or involving meat, the more regulations you will need to adhere to. You must do your processing in a commercial kitchen.

A much easier, accessible, and affordable option is to produce food products in your home kitchen under your state's cottage food law, which provides options to bypass commercial kitchen requirements — legally.

Shelf-stable value-added products like salsa or sourdough bread can, in many states, be processed in home kitchens and sold at venues like farmer's markets, particularly winter and holiday markets in the off-season. According

Cottage Food Laws in the United States

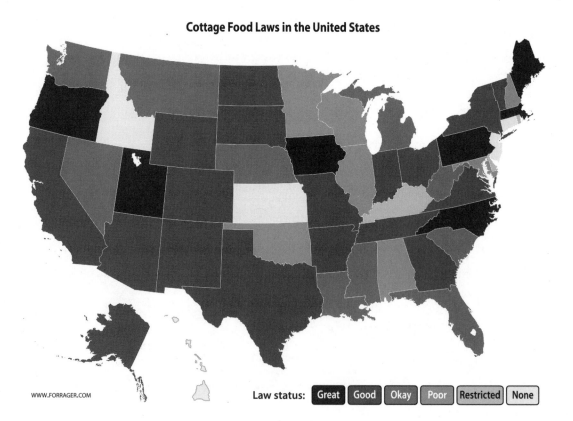

WWW.FORRAGER.COM

Law status: Great   Good   Okay   Poor   Restricted   None

to the USDA's *National Farmers' Market Directory*, winter markets — markets that operate at least once between November and March — have been increasing by over 50 percent annually in the past couple of years, now accounting for nearly 24 percent of the directory's 7,865 farmers' markets.

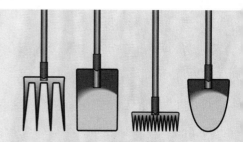

Tool Shed:

Homemade for Sale

"Homemade" and "fresh from the oven" on the package can mean exactly that. Over 40 states have laws known as "cottage food legislation," which encourage farmers and home-cooks to create and sell specific, "non-hazardous" food items, often defined as those that are high-acid, such as jams and jellies, or low moisture, like breads. Whether making pies or preserves, wedding cakes or granola, pickles or decorated cookies, anyone can diversify income by producing non-hazardous foods in the home kitchen.

John and I wrote *Homemade for Sale: How to Step Up and Market a Food Business From Your Home Kitchen* to support and champion folks to become food entrepreneurs quickly and cost effectively by launching a home business. Starting off under cottage food law enables you to test the business waters with practically no start-up expense, as everything you need is already in your home kitchen. If your product gets rave reviews and takes off, you'll be in a much better strategic and financial position to look into renting or even building your own commercial kitchen.

For more information on *Homemade for Sale*, visit www.homemadeforsale.com.

Also on the website is a free downloadable toolkit specifically covering labeling, packaging, and cost-efficient farmers' market displays for canned items in jars.

Regina Dlugokencky, Seedsower Farm, Long Island, New York

Fermenting Dough with Home-baked Bread

"I'm a small-scale nano-farmer looking to keep the income going in the winter," shares Regina Dlugokencky, owner of Seedsower Farm, a little market farm operation in Long Island, New York. What started as an afterthought to the farming season fermented into a bread-making obsession for this self-taught baker. "You could say I've built a nice little cult following for my bread," she adds. "I'm now known as 'the organic bread girl' at our farmers' market."

Seedsower Farm owner Regina Dlugokencky holding her fresh-baked bread.
Courtesy of Seedsower Farm

Dlugokencky sells sourdough breads and organic jams every Sunday from December through April at the Long Island Winter Farmers' Market in Huntington Station, New York. Using only organic ingredients and New York state flours, she quickly discovered the sweet spot in starting a bread business.

"The first time I experimented with bread at the market, I just brought six loaves. I used one for sampling and quickly sold the other five. I was hooked on the potential." Like many cottage food operators, Dlugokencky used the farmers' markets as "test markets" for her products, examining customer preferences and price sensitivity and gauging demand.

Today she bakes 30 to 40 loaves each week, selling for between $5 and $6.50. In the spring, Dlugokencky also raises organic vegetable plants. Her other growing-season focus is crops that will serve as additional products for the winter markets, such as berries for her jams and storable alliums like garlic, shallots, and specialty onions. She cures these to sell at the winter markets.

Blending efficiency and a dedication to frugality, Dlugokencky creates various systems and organizational techniques to turn out loaves from her small kitchen. "I didn't go out and buy everything right away when I started. Instead, I tapped into friends' offers to help. It amazed me how many folks owned multiple KitchenAid mixers and were happy to let me borrow theirs. "The day before market, she makes the dough, using a sourdough starter instead of yeast, which gives the bread that desirable tangy taste and crusty texture. She lets the mixer do the heavy work of kneading. The dough is then placed in large containers for the long, first rise.

After an initial rise of three to four hours, she divides the dough and weighs each piece on a scale for consistent loaf size, finding that her customers generally like a smaller loaf, (about a pound after baked). After a short rest time, she shapes the loaves, places them in bowls, covers them with plastic wrap, and sets them in the refrigerator for the final, slow rise overnight.

At 2 AM the next day, Dlugokencky's baking is in full swing. She preheats the oven and removes the loaves from the refrigerator to bring them to room temperature. To make a 10 AM market start, she bakes from 3 to 8:30 AM.

"My biggest challenge is baking all the bread in my kitchen in time for market. I . . . finally invested in a large 5.6-cubic-foot oven with convection. Now I can bake eight loaves at a time." She thoroughly researched models and corresponding reviews before deciding on a Kenmore that met her $1,000 budget. "I considered buying a commercial oven, but the electricity would have to be redone and, technically, a home processor's license in New York disallows use of professional equipment."

"At market, let your customers pick out which loaf they like," Dlugokencky suggests. "It's an easy way to add that personal touch and interact with them." She also makes a few batches of onion focaccia and sourdough popovers, which she sells individually. These add a tempting impulse purchase since they're something her customers can eat right away.

"I prepare jams from the summer berry harvest, but I like making bread better. The jam is a much smaller side business, less than 20 percent of my total sales," adds Dlugokencky. In addition to the berries she grows herself, she buys local, organic fruit in bulk in season and freezes it to make jams during the winter. She charges a $1.50 deposit for the jar to keep the costs down and finds that customers have no issue with that, even though she rarely gets the jar back. As a vendor at the Long Island Farmers' Market, Dlugokencky must carry both general and

Seedsower Farm jam.
COURTESY OF SEEDSOWER FARM

product liability insurance, which she gets through Farm Family Insurance for a couple of hundred dollars a year.

"Start small and keep learning," Dlugokencky advises. "Once you find out what you like to do, like me with bread, fine-tune your recipes and don't be afraid to fail. The first loaf of bread I made doesn't compare to what I can pump out in multiple batches every week now. It took practice and some mess-ups along the way."

Dlugokencky's frugal nature embraces and celebrates the failures: any bread not fit for sale makes great breadcrumbs or croutons!

At this writing, under Wisconsin's current cottage foods law, I can produce high-acid canned items such as jams, salsas, and pickles in our home kitchen for sale at farmers' markets or community events. After a bumper crop of cabbage and pumpkins one summer, I got into sauerkraut and pickled pumpkin, which ended up being distinct enough products that I sold out at our local holiday market.

But what I clicked with was the labeling and packing — which ultimately also boosted sales. It doesn't take the National Association of Wrapping Paper (if such a group exists!) to tell me what to do. I played with colorful ways to dress up the jars and make versions in Christmas colors for holiday sales. My trips to Michaels were deductable as a legitimate supply expense. No need to twist my arm to linger there! And those sales help bring in revenue year-round.

For some of us, keeping our businesses relatively small, in home kitchens and under cottage food law, will remain the best option. But maybe, like Dorothy Stainbrook of HeathGlen Farm & Kitchen with her Blueberry Lavender with Merlot Wine, you have a product that is not just great, it's awesome, and you're hitting your state's cottage food sales cap. Maybe you're hounded by specialty food stores that can't wait to stock your zucchini bread as soon as you can sell wholesale. On a personal level, perhaps you want to scale up your operations because you're having so much fun and like the thrill and excitement of running your own enterprise.

Scaling up moves you into the realm of commercial kitchens. Several options exist at this point, including renting space at an incubator community kitchen. This type of facility caters to small food business start-ups, but they are typically located in more urban areas. You may be able to use a local church kitchen, depending on their licensing and willingness to work with you, or rent a restaurant kitchen during off hours.

Another option is working with a contract packer, or co-packer, which is a company that processes food products, either using its surplus capacity or specializing in packing other businesses' food items. You turn over your recipe, perhaps some ingredients if you're growing them yourself, and any other marketing elements, and the co-packer takes it from there. What you get in return is a ready-to-sell product.

Depending on the co-packers, they might provide the following services:

• Production of the product
• Guidance on product formulation and development, including reformulating home recipes to large-scale production processes
• Packaging of the product
• Guidance on aspects of labeling, especially related to regulatory requirements
• Advice related to marketing and distribution

From a time and labor perspective, this route allows you to quickly and professionally produce your product during peak summertime. Due to the volume, this may afford greater product consistency and quality and economies of scale. Additionally, co-packers can assist with the requirements that arise as you move from being a small cottage-food enterprise to a wholesaler, including UPC codes, nutritional labeling, and any lab analysis that may be required.

Going with a co-packer comes with a higher per unit cost. You will need to evaluate this increase in terms of time, labor, and the financial resources required to produce your items another way. If this streamlined and complete process of production seems too good to be true, don't forget the price and minimum-order requirement.

For example, Peg Sheaffer at Sandhill Family Farms in Brodhead, Wisconsin, uses a local co-packer that specializes in items that are "pumped," such as sauces, to turn their extra tomatoes into canned organic pureed tomatoes. Although the final product they receive is labeled and ready for wholesale, Sheaffer chooses to use most of it to add value to her CSA boxes during the lean early spring. The key with co-packers is that they typically require a large minimum volume of produce to work with. Sheaffer's co-packer requires a minimum of 5,000 pounds of tomatoes.

An appealing option for many women, but also the most costly, involves building an on-farm commercial kitchen. This is what Stainbrook decided to do. (Read more in "How She Sows It.") While this venture can run into the tens of thousands of dollars, it enables you to remain on-farm and could potentially be used for other value-added enterprises, such as on-farm food service, which we'll talk about in the next chapter. With much higher price entry points, proceed slowly and with fiscal caution. *Homemade for Sale* includes a section on scaling up that examines these commercial options.

Co-packed tomatoes from Sandhill Family Farms.

SANDHILL FAMILY FARMS

Got some business idea seeds planted? While this chapter covers core farm business, read on to the next chapter, which is the advanced course of diversification, dialing into the more experiential and education-al side of farm life and how to add that in as an income source to your farm livelihood.

Dorothy Stainbrook, HeathGlen Farm & Kitchen, Forest Lake, Minnesota

Jamming Expansion:
Building an On-farm Commercial Kitchen

"Your best research comes directly from your customers. Ask them what they like, and make it," shares Dorothy Stainbrook, owner of HeathGlen Farm & Kitchen, specializing in preserves, syrups, and shrubs made from the organic fruit from her farm. Stainbrook lives the lesson behind this advice: She launched and tested her jams at farmers' markets in Minneapolis and St. Paul while operating under Minnesota's cottage food law. Thanks to customer feedback, Stainbrook found a niche in unique preserve flavor combos like blueberry lavender with merlot wine. With this sales success moving her closer to her state's gross sales cap, she expanded and invested in building an on-farm commercial kitchen.

To understand and learn from Stainbrook's evolution, jump back to 1998. "Our five-year-old daughter at the time had some medical problems that needed Mom to stay closer to home. As I gazed out the window at our 23 acres, even though those fallow fields were at the time overgrown with weeds, I saw the chance to trade my white-collar career and follow my true dream of starting a farm." HeathGlen organic farm evolved from this vision, five acres of primarily berries along with herbs and vegetables sold at the St. Paul and Mill City Farmers' Markets from May through December.

"Almost ten years later, in 2005, that same daughter decided some jellies would be nice for the peanut butter sandwiches she took to swim practice," recalls Stainbrook. "Turns out we had a bumper

HeathGlen owner Dorothy Stainbrook in her commercially licensed home kitchen.
COURTESY OF HEATHGLEN FARM & KITCHEN

crop of fruit that year, so I confess I went a little overboard trying all kinds of pepper, wine, and fruit jellies."

This value-added direction proved to be something Stainbrook could readily experiment with under Minnesota's cottage food law, and it strongly fit with her future business vision: "I wanted HeathGlen to be more than a hobby; I wanted to make a full living on the farm. I saw these fruit preserves as an opportunity to develop a part of the farm business that would take me through the whole year financially, especially the winter months."

"I did a ton of sampling at the market to get feedback from customers while developing a unique distinction by keeping the sugar as low as I could. This tremendously accents the fresh fruit flavor." Stainbrook also tapped into her former bartending expertise and blended liquors into the preserves to enhance the fruit flavor. Volume and sales snowballed to the point that Stainbrook exceeded the $5,000 Minnesota gross sales cap in 2008 and needed to look into commercial kitchen venues.

"I first rented commercial kitchen space at a local church that had a state license and had rented to businesses like mine before," explains Stainbrook. "Even though this kitchen space was just a mile away, I quickly realized the hassle in packing and lugging my ingredients and equipment back and forth from the rented space. I also hated being away from my kids, who were young at the time, and I also needed to be on-farm to supervise my staff of part-time employees."

After two years in the church kitchen, Stainbrook decided to build a commercial facility on her farm. "If you grow your own produce and don't have on-site storage at your rented kitchen venue — always my situation — building your own commercial kitchen may make sense from an efficiency standpoint." Stainbrook is quick to advise of the importance of patience and understanding the time and cost involved with building such a facility. Hers was the first on-farm facility that many of the state regulators had to work with, so they were often learning together on what, exactly, it needed to look like.

"I love what I have now, but there were lots of hurdles and expense to get here." It took her about two years to build her kitchen, including one year in the planning and permit stage. The kitchen was built within the existing attached garage but needed a new floor, walls, and lights to reach the Minnesota state code.

A reality of plowing new ground with state regulations is that HeathGlen Farm needed to comply with a litany of stipulations that were overkill for a small canned-food business. "To comply with different regulations, we definitely built more than we realistically needed. We had an additional handicap-accessible bathroom, even though I was the only one working here. A huge ventilation intake system covers a six-foot area over the stove, which is overkill for the size of the kitchen space." The commercial kitchen cost about $50,000, which was financed by family savings.

The new kitchen was a playground for product experimentation. With the growing interest among foodies in "craft cocktails," Stainbrook tapped into her bartending background and started making fruit-flavored syrups and shrubs, a vinegar-based fruit sugar syrup that dates back to colonial times. She sells these jarred, value-added products at both summer and winter farmers' markets. The winter sales average higher, at around two hundred jars per event. She charges eight dollars for a half-pint jar of preserves and for an eight-ounce bottle of syrup.

With the syrup and shrub additions, Stainbrook realized the importance of customer education to showcase how her products could be used. Her "Balcony Bartender Blog" features YouTube videos demonstrating how to use the products in craft cocktails. She also maintains a recipe blog that offers creative ways to incorporate her products into various dishes. All of this can be accessed on her website.

"The success of my preserve business led to the expansion and the advent of the new product lines of syrups and shrubs because I could process these at home," Stainbrook adds. "I'm thankful for cottage food laws. As a farmer, I didn't have the money or the time resources to go into the city and rent a commercial kitchen when I got started."

HeathGlen Farm & Kitchen syrups displayed at a farmers' market stand. COURTESY OF HEATHGLEN FARM & KITCHEN

Chapter 6

Complementary On-farm Enterprises

MAYA ANGELOU ONCE SAID, "You can't use up creativity. The more ideas you use, the more you have." Even though she never milked goats or sold pickles, this prolific poet's words sum up how we women farmers view our farm operations: a canvas of ideas. The wider we open our eyes and minds, the more business possibilities we see.

My farm stores a long list of business brainchildren. That old granary building with sideboards falling off? Perhaps I could infill it with straw bales and transform it into a solar-heated greenhouse. A small

Laundry blowing off the line at Inn Serendipity.

JOHN D. IVANKO

opening of empty pasture? I see labyrinth potential. My laundry blows off the line? I start to wonder if we have enough wind to erect a wind turbine and generate our electricity. That old milk house next to the dairy barn? Well, with its cement floors, drain and solid windows, I could readily upgrade it to a commercial kitchen with a wood-fired pizza oven for pizza on the farm.

If you're like me and have what feel like a million ideas bouncing around in your head, you're not alone. Diversification reigns supreme among us women farmers. There's strength in diversity. Mother Nature doesn't monocrop or plant one seed. Got proof: behold the apple tree in spring, buzzing with honeybees pollinating the fragrant flowers. The same holds true for business ideas. We don't see an old beat-up barn. With a little elbow grease, tulle, and tables, we can see weddings hosted on our land.

This chapter will help you synthesize and understand six key categories of farm-based businesses outside the production focus of your operation. First, let's dig a bit more into this concept of diversification and what it means to run a diversified farm business.

Diversify it: risk management

In farmer training venues, particularly via the USDA, you'll often hear the term "risk management" as something farmers should keep top of mind as we plan our businesses. We live in increasingly uncertain times. From Hurricane Katrina to Sandy, from extreme drought to flooding and spikes in energy costs, the only certainty is uncertainty. The impacts of climate change cloud even the most well-thought-out business plans with gloom. Every seed we plant is under the rule of Mother Nature, and she's increasingly pissed about what we're doing to her planet by throwing it out of kilter. Despite our heartfelt commitment to conservation and sustainable agriculture, we suffer these un-predictable climate consequences of the lack of stewardship. We're vulnerable and increasingly can share stories on how climate change affects us. At Inn Serendipity, our wind turbine, a Bergey 10kw system, suffered blade damage thanks to extreme straight-line winds.

Don't wallow in gloom or give up. We do have a solid strategy to manage our risk through these erratic consequences: diversification. Diversification = need + creativity + action. Diversification fits nicely under risk management because it creates more financial security. With various ventures going on, if one element isn't generating revenue, your business is still afloat because an-other area will be.

I learned the merits of diversification early on when I was new to the pro-duction side of farming. Our harvest yields were small and we had not yet established any local sales venues, but we had already started the bed and breakfast and had that income as a base. What we did grow became the foun-dation of our vegetarian breakfasts. Our zucchini harvests reigned in years two and three (don't ask about the first season when we embarrassingly couldn't germinate zucchini!) and appeared in multiple forms on breakfast plates.

This "working with what we had" strategy served the farm well as the breakfasts and the recipes we developed over the following decade became the basis for cookbooks, most recently *Farmstead Chef.*

This same strategy holds true with having multiple sales outlets. This follows Grandma's advice: "Don't put all your eggs in one basket." If you have only one wholesale account for your eggs and that place closes, you're up a creek without an eggbeater and need to build new markets fast. If, instead, some sales went to that store but you also sold at a farmers' market and to restaurants, your income would come from different sources and you'd still survive if one fell through.

Chocolate zucchini roll with raspberry sauce from Farmstead Chef.

JOHN D. IVANKO

Diversification also manages risk when it comes to generating seasonal income throughout the year. Let's say the majority of your income comes from produce sales during the summer growing season, but you need an influx of cash in the early spring to buy seeds and equipment. By doing some value-added products in your home kitchen, you could sell at winter farmers' markets and keep that income stream coming in.

We add the creative feminine to farming. As we create multiple businesses, ideas bounce off each other. This builds more than business financial muscle; we feel that inner satisfaction of making something better than what was. We leave our distinct mark, our trademark style, on all elements of our farm enterprises, playing by our own rules.

For example, tradition would dictate that a fall gathering for your CSA members would involve carving pumpkins. But that wasn't enough for Tara Smith of Tara Firma Farms in Petaluma, California, north of San Francisco. Hers

Pumpkins on Pikes at Tara Firma Farms.

is no ordinary Halloween party. Each carved pumpkin is placed on 4″ × 4″ wood platform nailed to a stake which can be three, six or nine feet in height. These stakes are pounded into the ground at varying heights throughout a field prior to the event. As dusk falls, a tea light is lit inside each pumpkin, and the valley setting glows with over 700 "pumpkins on pikes."

"When darkness falls, the pumpkins look like they are floating in the air. It is a magical, surreal experience," explains Smith. "Pumpkins on Pikes reflects my true passion of changing our food system as it provides a compelling reason for people, particularly families, to come out to the farm. They experience so much more when they are here, from building community with others to helping create this amazing field of beauty."

Smith streamlined logistics over the years: Biodegradable tablecloths cover long tables for carving and she provides simple carving kits and Halloween supplies for attendees who don't bring their own. Garbage cans collect the guts of the pumpkins, which end up as compost or chicken or pig feed.

Embrace the innate female ability to juggle, manage, prioritize, and cope. A nugget of advice from the thousands of women I've had the pleasure to interview: Keep it lean, and start slow. Out of my ideas I threw out in the beginning of this chapter — straw bale greenhouse, labyrinth, wind turbine, and pizza nights — two out of four happened, one died a weedy death, and one is still being incubated. The dilapidated granary evolved over several years into a retrofitted straw bale greenhouse that we use for germinating spring transplants and curing garlic and onions. Our son took over the first floor with his Lego and, later, computer tech lab.

The laundry blowing off the clothesline prompted a lengthy learning process about renewable energy systems and eventually the installation of a wind turbine that more than meets our electricity needs. The electricity overproduction comes back to my farm in the form of a credit check of $200 to $300 a year.

The labyrinth falls under the category of too long a winter that resulted in too many new ideas. I read several articles about labyrinths, illustrated with glossy color photos. I fell in love with the concept, one that dates back to ancient times. I thought this would be a fabulous addition to the farm as a meditative activity and would be a marketing draw for B&B guests.

I overlooked how much effort a labyrinth takes to maintain. It became more work and cost than benefit, and definitely not fun at all. After two seasons we let it go back to its natural state.

But here's a serendipitous twist: Because we disturbed the soil for the labyrinth project, maple trees popped up in earnest from the seeds of a nearby tree. My son reinvented this area into a shady nook filled with hammocks. My lesson in diversity: be open to something sprouting up that's different from what you planned — and better than you imagined.

The milk house conversion to pizza on the farm? Stay tuned. I'm mulling and researching slowly. We took step one and purchased a commercial NSF certified, outdoor wood-fired oven. For now, it's for purely personal use and I've been working my bread-baking techniques. We've also been hosting afternoons we call Community Open Hearth, when we invite local friends over and fire up the oven for folks to bake and hang out.

Sleep it: farm stays

For many women looking at multiple farm-based income sources, providing

Bergey wind turbine powering Inn Serendipity.
JOHN D. IVANKO

Straw bale greenhouse at Inn Serendipity.
JOHN D. IVANKO

Lisa walks the labyrinth at Inn Serendipity.
JOHN D. IVANKO

Hammock-ville at Inn Serendipity. John D. Ivanko

Baking in the outdoor wood-fired oven at Inn Serendipity. John D. Ivanko

hospitality via a farm-stay experience holds strong appeal.

On most summer mornings, follow the aroma of cinnamon and nutmeg and you'll find me in the kitchen baking muffins for B&B guests, using squash puree I froze last fall. The sun rises as I'm in the garden, harvesting an armful of Swiss chard and dill for that morning's quiche. When I host guests who like to sleep in a little later, the sun will be high enough that I can take out my SunOven and bake that morning's egg pie with solar rays.

Eventually I hear footsteps coming down the staircase from the bedrooms as guests arrive to eat. By this time John is also up, and conversations perk up as the coffee flows. Technically "strangers" last night when they came up our driveway, our guests share life tales around my kitchen table, and we soon feel like old friends.

Each state determines the requirements for B&Bs. Check with your state's health department or B&B association. Typically the requirements are basic safety elements, such as smoke detectors and fire extinguishers, and do not require you to have a commercial kitchen. As Airbnb grows in popularity as a way to advertise your business, remember it is completely separate from your state requirements. You need to still officially go through your state's regulations and requirements to legally book reservations through Airbnb.

There are two questions I occasionally get from women interested in starting a B&B: "Are you ever worried someone will steal something?" and "Don't

Lisa bakes with the Sun Oven. JOHN D. IVANKO

you miss your privacy?" My answers are "no" and "no." I tell them straight out that, honestly, the fact that they even ask those questions waves a red flag that the B&B lifestyle is not for them. There's an immediate level of intimacy and trust that unfolds when strangers move in to spend the night. If you're on guard or remotely concerned of theft — or feeling the need to be by yourself — the farm stay simply isn't the lifestyle for you. That is nothing negative, and definitely no judgment made. It's important to ask questions, realizing sometimes we'll hear answers that cause us to question ourselves and explore other venues.

Idea Seed:

"Accidents on the farm with guests tend to happen in the first thirty minutes in unfamiliar territory, so make sure you take them on a tour when they first arrive, showing them where they can and cannot be, what to watch for, and what the rules are. Also, guests who sleep well are happy guests. Spend the extra bucks on good linens, mattresses, and pillows."

— Scottie Jones, Founder of Farm Stay U.S. and owner of Leaping Lamb Farm, Alsea, Oregon

Farm-stay ventures can often directly support your other businesses, such as produce sales. I speak from 20 years of experience when I say make it easy for guests to know what you have for sale and the prices. For the first 12 years

Inn Serendipity®
7843 County P • Browntown WI 53522
Tel: 608-329-7056 • E-mail: info@innserendipity.com
www.innserendipity.com

Take Home a Taste of the Gardens at Inn Serendipity
Savor the good life with our organic produce, harvested just for you. Use this form to place an order for seasonal, farm-fresh produce currently available, HIGHLIGHTED below. Please place your order the afternoon or early evening before check-out so we can do an early morning harvest for you. Thank you.

	Size	Price	Amt. Ordered
Asparagus:	pound	$7.50	
Basil:	5 oz.	$2.50	
Beets (gold, red, chioggia):	pound	$3.50	
Broccoli florets:	pound	$3.50	
Broccoli head:	pound	$4.75	
Cabbage:	pound	$1.50	
Chives/wild onions/ramps	huge bunch	$2.00	
Cucumbers (eating/pickling):	pound	$3.50	
Dill:	medium bunch	$2.25	
Garlic:	pound	$8.50	
Green Beans:	pound	$3.50	
Kale	lg. bunch	$2.50	
Leeks (large):	pound	$2.25	
Melon (musk./waterm.):	pound	$1.75	
Onions (white/yellow)	pound	$1.75	
Peppers (green Bell):	pound	$4.75	
Potatoes (red/gold/russet):	pound	$1.75	
Pumpkin (cooking varieties):	pound	$3.25	
Radish (red)	bunch (4-6)	$2.50	
Rhubarb:	pound	$3.50	
Salad mix/lettuce:	lg. pack	$2.25	
Spinach (perpetual):	lg. pack	$4.50	
Strawberries:	pound	$4.75	
Swiss Chard mix:	lg. pack	$4.50	
Peas (Sugarsnap/Snow):	pound	$5.75	
Tomatoes (red/green):	pound	$3.75	
Winter Squash (carnival/ butternut)	pd.	$3.25	
Zucchini (8-ball/pattypan/curly):	pound	$2.75	
Flower bouquet	**jumbo**	**$5**	
Home-canned Sauerkraut, quart/pint		**$8/$6**	
Home-canned Dill Pickles, quart/pint		**$8/$6**	
Home-canned Sweet Pickles, qrt/pint		**$8/$6**	
Home-canned Relish	**half pint**	**$5**	

of Inn Serendipity, we would casually mention to guests during a welcome tour that we offer fruits and vegetables for sale. In those 12 years, we sold about 12 bucks of produce. Whoops.

In year 13, we placed an easy-to-use order form in guests' rooms. The list includes everything we grow, with pricing based on what Willy Street Co-op in Madison, Wisconsin, charges, and I use a highlighter to mark what is currently available. Guests place orders and we can custom harvest. By making our produce obvious and "special" with the idea that we "custom harvest," we jumped to average sales of $15 to $50 per stay, especially to guests who don't belong to a regular CSA or are not backyard growers themselves.

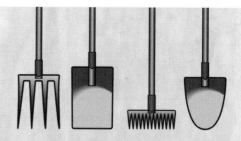

Tool Shed:

Start-up Resources for Farm Stays

Farm Stay U.S.

www.farmstayus.com

A national directory to advertise farm-stay operations (various membership levels available), they offer a free "Farmers Guide" with farm-stay start-up resources and advice.

Diversifying Your Farm Business Through Agritourism: A How-to Manual for Establishing a Farm Stay in Minnesota

http://www.misa.umn.edu/

A free download from the Minnesota Institute for Sustainable Agriculture. Though the regulations are specific to Minnesota, much of the business planning advice is applicable anywhere.

Gabriele Marewski, Paradise Farms, Homestead, Florida

Sowing Synchronicity

"This is art," proclaims Gabriele Marewski as she points to a red giant hibiscus, with its velvety petals so temptingly soft you want to stroke them. But then again, this is a farm, a fact Marewski quickly reminds you of as she picks the flower and pops it into her mouth. "Divine," she remarks, eyes closed.

Marewski has nurtured Paradise Farms on five acres in southern Florida over the past 16 years, turning an abandoned and overgrown avocado orchard into a blooming tropical tapestry. Befitting its name, Paradise Farms is a thriving ecosystem, a cacophony of birds seeking their refuge amid the mangos, star fruits, and avocados. "It's in my creative nature to have a lot of different elements going on," Marewski admits.

That artistic spunk drives the diversity of business ventures, contributing to the success of Paradise Farms. This includes raising specialty crops such as microgreens, baby greens, oyster mushrooms, and over 50 varietals of edible flowers, scheduled for immediate delivery to high-end chefs in Miami. Her latest business addition includes a bed and breakfast and farm-to-table dinners. The dinners are catered by top chefs she's developed relationships with over the years.

Marewski never concerned herself with how to fit into the male-dominated world of agriculture. Rather, she focused on creating her own niche by celebrating the feminine. "Every door on my farm is painted pink, simply because I can," explains Marewski,

Gabriele Marewski of Paradise Farms.
JOHN D. IVANKO

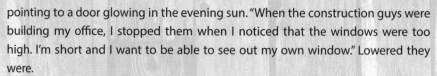

pointing to a door glowing in the evening sun. "When the construction guys were building my office, I stopped them when I noticed that the windows were too high. I'm short and I want to be able to see out my own window." Lowered they were.

"By being in touch with your feminine side, however you define that, you can express yourself and importantly be yourself in your farm business," Gabriele advocates. "For example, I wear skirts every day. This may not be typical farm gear, but it's who I am and what I like."

She also embraces her feminine side to nurture and care for her workers, of whom over 90 percent are women. "I started the oyster mushroom operation so I could keep my staff on the payroll year-round, not only during the high season. It helped my bottom line plus took care of the people important to me, my staff." The school bus drops their children off after school.

The microgreens business also evolved as homage to the feminine. "Women naturally like handwork and not necessarily operating large machinery. These microgreens are managed by hand, cut and packaged two times per week," she adds. Marewski uses cafeteria trays to grow her greens because they're easy for women to handle. Holes are drilled in them for drainage.

"It's about synchronicity and intuition. If you're connected to who you are as a woman and what your passion is, you're more aware of the world around you," Marewski explains. "Then when an opportunity opens up, you're ready to embrace it, even if you're not sure at the time exactly how the pieces will fit together."

That was the case with the "cooker" dehydrator, as she calls it. Someone had a large industrial cooker valued at $20,000 but basically giving it away just for the delivery price of $500. "I had no idea what we'd use it for, but it seemed too good to pass by." She later converted the cooker into a dehydrator to dry flowers and herbs for tea.

The farm dinners, called "Dinners in Paradise" opened the farm to the public. Guests dine amid the farm's blooms, fragrances, and hues in an open-air gazebo. "We custom harvest orders for the chefs who prep that night's meal. They use their restaurant

Garden shot, Paradise Farms.
John D. Ivanko

kitchen to prepare the meal and bring the pre-prepped food to serve on the farm as if they were a caterer," explains Marewski. This system works well for the high-end dinners, allowing Marewski to avoid having to invest in an on-site commercial kitchen.

The farm stays are in simple bungalows scattered throughout the property, particularly attractive to dinner guests who don't want to drive back to the city after dinner. Marewski offers guests a menu of options to slow down and relax through reconnecting with nature, from showering outdoors under the stars to dangling tired feet in a natural pool. "I love hosting folks on the farm, and people were asking about spending the night, so adding the farm stay proved to be a natural fit."

Her biggest challenge was running the farm as a single mom after her divorce. "I didn't have a partner to bounce ideas off," shares Marewski. "My son, Max, was in middle school at the time and I homeschooled him for a couple of years.... He helped in researching key business decisions, like what mower we should get. He recommended a more expensive option because it was a local business who would stand by their product. That's my boy!" For the past four years, Marewski, now entering her sixties, has had a farm manager, Landy.

Marewski's latest venture is writing down her vast encyclopedic knowledge and practical know-how in an easy-to-use binder format book, perfect for begin-

Farm-stay bungalows at Paradise Farms.
JOHN D. IVANKO

ning farmers, who can customize it for their own operations. "All someone has to do is follow along and use their own information. I wanted to create a training roadmap for beginning farmers to keep their own operation focused and organized."

"I've invested every dollar I made back into this place, growing slowly and having enough equity to take on new directions when timing and synchronicity are right. And it turned out better than if I put all my money in the stock market."

Taste it: farm-to-table meals

In my "In Her Boots" workshops, almost every beginning female farmer I talk to has dreams of doing some form of farm-to-table meals. What's not to love: enthusiastic people dining on our farm under the wide rural sky, savoring specialties made with our fresh fare, and connecting directly with the farmer who raised it, us? No wonder farm-to-table dining is hot, both among diners and farmer hosts.

But I'm always the bad cop who has to tell the enthusiastic fledgling farmer that, simply put, it's not that simple. The reality is you can't simply cook up food in your home kitchen, put it on a plate, serve it, and charge for it. Charging money for prepared food activates complex regulation and state inspector involvement. Like McDonald's and that family-run diner in town, you're serving food to the public and need to follow very specific requirements as to how that food is handled at all stages. This includes everything from needing commercial-grade equipment to washable walls and floors.

Point of clarification: We're talking about food service now, not producing food products. While your department of agriculture typically regulates food products, your state's department of health inspects and handles food service, including restaurants and catering operations. Food products and food service each have a completely different set of regulations. I can process jars of pickles in my home kitchen under my state's cottage food law and sell them to you at specific venues in Wisconsin. But once I crack open that jar, stick a pickle on a skewer, and sell it you like that, ready-to-eat, it's a completely different ball game, with a dizzying array of more stringent paperwork and permits.

This section aims to help you navigate the basics of on-farm food service, including outlining simpler, cost-effective options to test the business waters.

"While farm-to-table may be increasingly of interest as a business idea to farmers, particularly women, we identified a lack of resources specifically supporting successful start-ups," explains Jan Joannides of Renewing

Edible Madison's fall dinner at Primrose Valley Farm.
JOHN D. IVANKO

the Countryside. Joannides and I convened a team of experts to research and create such a resource, which resulted in a new training program, "Come & Get It: What You Need to Know to Serve Food on Your Farm."

"Our goal focuses on empowering farmers with information before they start calling up state health departments, and to research a solid business plan before investing. Knowledge cooks up confidence and success," adds Joannides. The insight in this section is rooted in our learning in producing "Come & Get It," a collaborative effort of Renewing the Countryside, the Minnesota Institute for Sustainable Agriculture, Farm Commons, and yours truly, with support from the Minnesota Institute for Sustainable Agriculture Information Exchange and USDA Farmers' Market Promotion.

Let's start with the simplest option: the good ol' potluck. For the legal record, gathering together voluntarily and sharing food made in home kitchens is, in most states, now allowed under the law. Holy hot dish! Do you mean for years we've been cross-pollinating illegally over casseroles? Apparently so. In recent years, states increasingly have passed variations on "potluck laws," which specifically state that such gatherings are legal. The impetus for these laws often was a reaction to state inspectors showing up in the church basement at Grandpa's funeral lunch and sticking thermometers in the meatballs. That didn't go over well, so here we are with legislation.

Don't quickly write off the potluck as "been there done that" and not related to your farm business plan. A potluck remains the easiest and most readily accessible way for people to gather and eat on your farm. Income-generation aside, it meets many of your outreach goals of getting your farm name out there, bringing people together in community, and educating about where food comes from.

Another option to serve food on your farm is to partner with an event planner or restaurant familiar with these types of events. Functioning as a caterer, these outside entities bring prepared food from an off-site commercial kitchen. Working with such partners brings the event out of potluck status because attendees pay for prepared food. You could invite your CSA members to purchase tickets to savor a special night at your place. Their friends could buy tickets and come. Working with an outside group involves compensating their time to prepare the food as well as the resources they use, like commercial-grade equipment. You can still sell the chef ingredients from your farm and generate income that way, but you most likely won't be making much, if any, money off the ticket price with this kind of setup.

Come and get it!
DO I NEED A LICENSE?
Before you ring the dinner bell - Let's discuss what you want to do.

BRETT OLSON,
RENEWING THE COUNTRYSIDE.

When you move into food that you prepare yourself, your menu choice has a large impact on the degree of regulations, permits, and overall expense you will need to deal with, explains Rachel Armstrong, lawyer and founder of Farm Commons, a nonprofit organization that empowers farmers through legal education. Bottom line, you will not be able to use your home kitchen for any food preparation; you will need some form of licensed facility, whether it be a commercial kitchen, a food stand, or other option. What exactly is required to license the facility will be dictated by what you serve.

"A menu of re-heated hot dogs with a bag of chips, packaged cookie, and a can of soda will be much simpler in requirements than a plated full-course meal made with your vegetables, meat, and other farm fare," adds Armstrong. "Understandably, that isn't the most appealing option to farmers who are about local and fresh, but it's important to understand that what you serve makes the difference in determining what kitchen you need to invest in."

Cue the pizza farm as an accessible, intriguing, tasty option. With

roots originating in the Midwest, the "pizza farm" concept involves a farm serving these cheesy tomato pies, typically with farm-raised fare as ingredients and baked in an extremely hot, 800-degree, wood-fired oven.

Pizza Night on Friday night at Stoney Acres Farm. JOHN D. IVANKO

"The pizza-making concept is simpler from a regulatory standpoint than a five-course, multiple dish meal, and it brilliantly can still be a way for a farm to showcase its fare through topping combinations," explains Armstrong. Pizza farms serve the pizza "take-out" style, and guests have the option to take it home or, much more likely, bring their own gear and eat picnic-style on the farm. This format often simplifies life, as a pizza farm might be able to avoid infrastructure for things they don't do, such as serve food on plates or reheat prepared dishes like a full-scale, sit-down restaurant.

Typically farms run a pizza night once a week, which will often keep things simple under state regulation code that deals with frequency. Once you get past a certain number, more requirements kick in that push you into a more restaurant-like health code, which gets more complicated and costly, such as needing to test your water weekly.

"As you move beyond something like pizza into more complex menus, increased infrastructure and cost to your kitchen set-up will be required," Armstrong sums up. "Your best bet is to develop a strong and transparent working relationship with your inspector so you can plan your menu and overall business venture in a way that meets requirements and satisfies your food service vision."

Kat Becker, Stoney Acres Farm, Athens, Wisconsin

Pizza Adds Diversification to Farm Income Menu

Kat Becker pulling pizza out of oven at Stoney Acres.

JOHN D. IVANKO

"Diversification inspires us to continually be creatively challenging ourselves, thinking proactively about the future and always asking 'What if … ? '" explains Kat Becker, co-owner with her husband, Tony Schultz, of Stoney Acres Farm. Their certified organic operation is located about 30 miles west of Wausau in North Central Wisconsin. "Doing a weekly, on-farm pizza night proved to be one of our most lucrative and fun ventures yet. It brings together the ultimate combination for us: sharing what we grow and raise directly with our community, right on our land."

Now in their eighth season of farm production, Becker and Schultz run a highly diversified operation and serve as the third generation of farmers on Schultz's family land. The core of Stoney Acres Farm includes a 20-week CSA vegetable operation, along with herb, fruit, and flower production; grass-fed beef, pastured pork, and chicken; organic grains; maple syrup; and their newest venture, which officially opened in 2012, farm-to-table pizzas served on Friday nights from May through November.

"Diversifying into pizza made strategic sense on multiple levels as we already raised or grew most of the key ingredients, from pigs for the sausage to vegetables for toppings," explains Becker. Their key pizza cost is cheese, which adds up to $2,000 annually and is purchased direct from regional cheese-makers. "We saw the growing interest in pizza farms in other parts of Wisconsin and Minnesota and knew we could take advantage of being the first such venture in our north-central part of the state."

For those starting on the pizza farm business journey, Becker offers five slices of advice:

1. **Take time for research and planning**

   Stoney Acres' pizza operation showcases the importance of researching and planning strategically when diversification requires investment. "Installing a commercial kitchen isn't as intimidating as you think, but we did need to research and understand the requirements so as to use our money wisely," advises Becker.

   The commercial kitchen was part of a granary remodel and included purchasing kitchen equipment, $2,000 at auction, and installing washable walls. Their total cost was around $5,000, doing much of the labor themselves.

   "Take the time to visit other operations and learn how others operate," Becker continues. "We gained much insight from visiting other farms and seeing how they run."

2. **Collaborate and ask questions**

   "Most state inspectors and agency folks are on your side and want your business to succeed. But they have their rulebooks so that proper and safe food-handling procedures are carried out. You need to fit into their boxes," advises Becker. "Keeping dialog open and transparent from the start helped us develop strong working relationships with our local inspectors. We started talking way before we broke ground or spent anything, so we'd be on the same page."

   However, if specific rules and requirements don't make sense to you and your situation, don't be intimidated to ask questions. "Remember, the inspectors are just following their checklists and are not empowered to change anything. If you want something different, you'll need to go higher up the agency channel, ask for an exemption, and get it in writing," Becker offers.

   Such was the case for Stoney Acres and cooking sausage. The state's food service code called for an exhaust vent over the stove when cooking meat, which would have cost over $10,000. Given that they are operating only one night a week during the summer and spending only a few hours cooking the meat, Stoney Acres contacted the head state inspector and asked for an exemption, which he granted. Once they showed the local inspector that official exemption, the whole issue went away. "We had to initiate the exemption process and ask," Becker adds.

3. Keep evolving

The pizza business keeps growing: In 2013, Stoney Acres sold over $30,000 in pizzas at $18 to $20 a pizza and gross sales keep growing. "We realized after some super-high-volume nights during which we had to refund money because folks were waiting over an hour for their pizza, that we needed a second oven to keep up with demand, which we added mid-season this year. That made a huge difference immediately," Becker explains. They're still trying to figure out the best work flow and how much staff they need to best handle peak-season nights when they'll be pumping out way over 100 pizzas.

To further diversify income, Stoney Acres sets up a farmers' market stand next to the spot where attendees order pizza, which adds up to a couple of hundred dollars in sales per event. "We sell at a local farmers' market the next day, on Saturday morning, so our produce is already harvested, and it's easy to set up a small market table at pizza night," shares Becker. "The market stand also helps us visually explain a certain topping item that folks may be unfamiliar with, like a garlic scape."

Kat Becker and family at Stoney Acres Farm.

John D. Ivanko

4. Be true to your values

With sustainability driving Stoney Acres, Becker continually makes decisions with environmental values in mind. Pizzas are served on reusable pans. Cardboard boxes are provided for take-out, and you won't find Styrofoam anywhere. Compost buckets collect food scraps for pig feed. Stoney Acres provides water for free but doesn't sell any other beverages, including soda. "I don't believe in soda," says Becker with a laugh. "You can bring it if you want, but we're providing good old water."

Running a family-friendly business is also an important value of Stoney Acres. Their three young kids are a part of the pizza-night scene, under the watchful eye of grandparents while Becker and Schultz work. Five-year-old Riley already embraces the family's entrepreneurial spirit: He harvests sunflowers and sells them to guests for a dollar a stem.

5. **Prioritize your core customers**

"The core of Stoney Acres Farm remains our CSA, and we are fully committed to our members," explains Becker. "These families form the backbone of our operation and believe in what this farm stands for, and they support us through the ups and downs of small-scale agriculture." This group also makes up the core marketing for pizza night, primarily growing the business through word of mouth.

The pizza-farm venture reaches beyond an income source for Stoney Acres, though. It's a coming together of everything they value. "We believe in creating a family farm that serves our local community, moving toward environmental sustainability while providing a beautiful and constructive setting to raise a family," explains Becker. "When we see people enjoying our pizza as the sun sets on our land and kids are running around catching fireflies, it comes together for us and is so incredibly rewarding."

Tool Shed:

Come and Get It — What You Need to Know to Serve Food on Your Farm

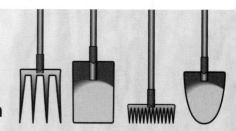

For more on setting up your on-farm food service venture, check out the "Come and Get It" training program. You can download a free webinar and manual at http://www.renewingthecountryside. org/on_farm_food_service.

While tailored for Minnesota and Wisconsin farms, much of the general information, from business structure to marketing ideas, is applicable wherever you may be. Farmers can attend on-farm field days at successful pizza farms and farm-to-table ventures to get a "behind the scenes" look at how it's done.

Experience it: agritourism

Agritourism is the broader umbrella category where hospitality meets farm life. While agritourism covers farm stays and farm-to-table meals, it also includes a variety of other farm experiences as diverse as corn mazes, U-pick pumpkin patches, and Christmas tree farms. Two trending agritourism ideas to consider include the following:

Glamping

A combination of *glamorous* and *camping*, "glamping" adds upscale sparkle and style to the pup tents and Coleman cookstoves of yore. From walled-in canvas tents to yurts, these accommodations provide an outdoor experience with the comforts of home, or at least a bed. But don't think that because guests camp this needs to be budget-priced lodging. Glamping spots can easily rent for over one hundred dollars a night.

Mona Campbell runs a glamping operation from Campbell Farm, nestled in the Laurel Highlands of Pennsylvania. Accommodations include a 12´ × 14´ platform tent, where guests snuggle in for the night on a queen-size bed complete with organic cotton sheets. The kitchen consists of a small refrigerator,

Tool Shed:

Glean More on Glamping

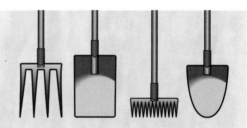

Glampinghub.com

This is an international directory of various takes on glamping accommodations, including barn, yurt, and campervan lodging you'd be more likely to find on-farm. But it's easy to get distracted looking at igloos, tree houses, and caves! Our Inn Serendipity Woods Cabin place is listed, too.

Glamping with MaryJane, by MaryJane Butters

If you're intrigued by glamping, check out this book. It will visually tour you through the art of Airstream renovation, cast iron cooking over a wood fire, and turning fishing lures into earnings. Butters so champions the blending of female grit and glam that she launched International Glamping Weekend, held annually the first week in June. Check out www.internationalglamping weekend.com to find gatherings near you.

electric lights, three-burner gas stove, double sink, hot and cold running water, and utensils you'd need to prepare meals — all under a roof but otherwise out-of-doors.

Weddings

Plug "farm wedding" into Pinterest and you'll see why the "rustic chic" look is so hip, simple yet stunning. From cocktails in canning jars to twinkle lights strung from barn rafters, farms provide the perfect backdrop for creative types who want to craft their own laid-back yet handcrafted style.

Glamping at Campbell Farm.

JOHN D. IVANKO

Teach it: workshops

Do you know how to can tomatoes? Spin fiber? Graft apple trees? Make goat's milk soap? Whatever your interests, there's likely a market of people wanting to pay you for your knowledge in a workshop setting. Knowledge and experience can add up to valuable workshop content. During the summer we host a Saturday afternoon "Sustainable Living Simplified" class about once a month. We charge 30 dollars per person for the three-hour class (a 10-dollar discount for B&B guests) and take folks through a detailed tour of the farm, growing fields, green-design renovations, and renewable-energy systems, answering any questions they may have. People often call and ask for a tour, and these workshops give folks an option that addresses their questions and interests and still adds to our bottom line.

Write it: articles and books

"Farming and writing can blend very well together as livelihoods, especially when you write about your first-hand, valuable experience and insight," shares Lynn Byczynski, publisher of *Growing for Market*, a national periodical for small-scale, ecologically minded farmers growing vegetables, fruits, herbs, and flowers for local markets, and a seasoned grower herself. "*Growing for Market* was born one spring day while I was picking peas for the farmers' market and

I wished I had someone to talk to." Building on the journalism career she had prior to farming, Byczynski launched *Growing for Market* in 1992 and it continues strong today, with ten issues annually.

Byczynski speaks from the heart when she encourages farmers to write what they know; her publication is exclusively written by market farmers, market managers, and others directly involved in the business of growing and selling local food and flowers. "Writing about what you know can be exceptionally efficient because you already know the information and don't need to spend eighty percent of your time researching the topic as the average journalist would."

The challenge in today's Internet-driven informational freebie world is getting paid for your words. "While the freelance landscape has changed over the last decade, I do see the pendulum shifting the other way now, back to paid content as publications and organizations realize that's how they get quality work," adds Byczynski.

"As writers, it is up to us all to demand to be paid our worth and be compensated," adds Anna Thomas Bates, a freelance food writer and co-owner of Landmark Creamery, an award-winning sheep-milk cheesemaking operation in southern Wisconsin. "If you're also doing elements like photography and recipe development and testing, those activities should be included in your payment. Reach out to writers doing similar projects to get an idea of what reasonable fees are for blog posts, newsletter articles, and feature writing so you know what to ask for."

Idea Seed:

"Everything in your life as a farmer is potential material for your life as a writer. Write things down. Seemingly unrelated events will start grouping themselves into a story, much like sheep gather themselves into a flock."

— Catherine Friend, author of *Hit By a Farm: How I Learned to Stop Worrying and Love the Barn* and *Sheepish: Two Women, Fifty Sheep, and Enough Wool to Save the Planet*

Work it: independent contract projects

When we moved to our five rural acres, the goal was — and still is — to generate our income from the farmstead but not necessarily through selling vegetables, fruits, flowers, and herbs exclusively. At first this stemmed from the simple fact that we didn't know much about growing food. You might recall that in our first year the zucchini laughed at us and never germinated. While income diversity has always been a priority, the underlying goal is to avoid having to work an off-the-farm job. Ever. No commute, no working for someone else, no wearing pantyhose and heels.

This approach to farm-based income opens the door to self-employed, independent contract work on specific projects for various organizations, but not as salaried staff. According to the IRS, "An individual is an independent contractor if the payer has the right to control or direct only the result of the work and not what will be done and how it will be done."

That works for me. Don't tell me how or when to do something; tell me what needs to be done and I'll make sure it happens. I operate in an independent contractor capacity for the nonprofit organizations I work for, like MOSES and Renewing the Countryside. This situation works out great for nonprofits, often strapped for cash and lacking the capacity to hire staff, which involves other financial obligations such as employment tax, health insurance, and retirement plans. I don't want any of that anyway; give me freedom to work on my own terms for organizations whose values and mission I support.

Again, according to the IRS: "You are not an independent contractor if you perform services that can be controlled by an employer (what will be done and how it will be done). This applies even if you are given freedom of action. What matters is that the employer has the legal right to control the details of how the services are performed."

As an independent contractor you are responsible for your own tax liability. Self-employed individuals must pay self-employment tax (SE tax) as well as income tax. SE tax is a Social Security and Medicare tax primarily for individuals who work for themselves, which is figured on the IRS Form 1040. It's similar to the Social Security and Medicare taxes withheld from the pay of most wage earners.

Your head is probably buzzing with business ideas by now. Excellent! Welcome to the world of women farming, where visions multiply like zucchini. Research more; talk to other women running such ventures; sleep on it; and mull some more. Trust me, there will always be more ideas than hours in the day. As we move forward into the pragmatic how-to behind making your livelihood a solid reality, start focusing on the core elements of your business plan and don't spin your wheels chasing every shiny new idea.

Chapter 7

Business-planning Boot Camp

OPERATING A WOMAN-OWNED farm enterprise isn't just about muddy boots left on a rack by the back door of the farmhouse, fresh produce by the crateful in the Coolbot cooler, and community potlucks on Saturday night. Like any successful business, there's plenty of thought and planning involved, including how to structure the enterprise, preparation of a business plan, and the mindful selection of technology that helps, not distracts, you from the farm chores and strengthening relationships with your customers. Without your customers, you won't be in business for long.

Boots on homemade rack by back door of Cathy Linn-Thortenson, Wise Acres Farm, Indian Trail, North Carolina.

JOHN D. IVANKO

This chapter delves into some of the basics of setting up your business and writing a business plan that includes a thorough marketing section. As the saying goes, if you fail to plan, plan on failing. Getting a plan is essential to success, but avoid over-thinking your business and holding off starting a farm until you have every last detail sorted out. News flash: you won't. Plus, politics change, people change, situations change. Nevertheless, setting up your business and writing a business plan can be both fun and extremely useful in evaluating every aspect of your business. The topic is covered in greater detail in *ECOpreneuring* and, related to cottage foods, *Homemade for Sale*.

Plan it: business planning

The practical aspects of starting up a business tend to be largely absent in any core educational program of higher education. Instead, we're instructed to learn a second language, take a course on fine literature or, for those in business school, suffer through "Organizational Behavior and Industrial Relations." The end game, too often, is to get a job.

While some colleges and universities are changing, albeit at the pace of creeping glaciers, getting a farm or farm-related business started doesn't require a PhD, or even an undergraduate degree. More important is an idea, commitment, perseverance, a bit of start-up capital, and some hard work. As alluded to earlier, some farms have funded their operations through crowd sourcing, CSA member shares, and other means, beyond turning to family, friends, or the bank.

Regardless of your ambition, you'll need to set aside seed catalogs to get down to the nitty-gritty of setting up your business, structuring it to serve your appetite for risk, and typing up a business plan that forces you to think through your ideas before you start unloading sheep, llamas, or pigs from the trailer.

Structuring your business

There are several ways to set up your farm business. You'll need to evaluate your own situation, your tolerance for risk, and your preferred type of business to determine which works best for you.

Below are several common business structures, broken down by the most recognized reason for choosing one over another: personal liability protection. This is a shield that prevents anyone with a court judgment or financial claims against the business from touching anything other than the assets of the corporation or limited liability company (LLC). In other words, certain business structures protect the personal assets of the officers, stockholders, and employees, reducing the risk that your house, personal property, or bank accounts could be seized as a part of a court settlement.

No personal liability protection

Sole Proprietorship

By far the most common, easiest, and least costly business structure is sole proprietorship or self-employed sole owner. Income from the business is reported as a part of the owner's personal income, using the IRS Schedule C or Schedule C-EZ; you may be subject to self-employment taxes of 15.3 percent. You are responsible for the liabilities and debts of the business. If your business is sued, everything you own could be threatened by the lawsuit.

General Partnership

If you go into business with a friend, sibling, or other family member, you would do so as a partnership instead of as a sole proprietor. When two or more individuals own a for-profit business, typically operating under a written partnership agreement, the business is a general partnership. All partners are responsible for the liabilities and debts of the business. Income is reported on the IRS Schedule K-1 and may be subject to 15.3 percent self-employment tax.

Tool Shed:

Dissecting Schedule F

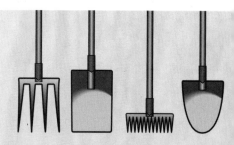

Schedule F is a profit-and-loss statement of your farm that is attached to your federal income tax return if you're a self-employed farmer (sole proprietor). The IRS defines a farmer as anyone who breeds fish, operates a ranch, grows crops, or raises livestock. For reference, the USDA defines a farm as "any place from which $1,000 or more of agriculture products were produced and sold, or normally would have been sold, during the year." Besides revenues and expenses associated with your farm business, you'll need to report any federal disaster payments, cooperative member royalties, and agricultural program payments, such as the Conservation Reserve Program (CRP), you receive.

Schedule F computes your net farming loss or profit, which is then reported on a designated line of your personal income tax return, known as Form 1040. If you made a farm profit, it will be combined with your other sources of income, perhaps an off-farm job, interest revenue, or stock dividends, increasing your total taxable income. Note, however, that farmers and fishers are required to pay their estimated tax for the previous year by January 15, using Form 1040-ES. According to TaxACT: "You have until April 15 to file your 2014 income tax return (Form 1040). If you do not pay your estimated tax by January 15, you must file your 2014 return and pay any tax due by March 2, 2015, to avoid an estimated tax penalty."

The partnership must file an annual return, Form 1065, with the federal government and possibly a state return.

Distinct legal entity offering personal liability protection to shareholders

S Corporation, or Subchapter S Corporation

Essentially a tax accounting classification, an S corporation is a common-stock-issuing legal entity, income from which is taxed only once, when it passes through to the employees or shareholders on their personal income tax returns. Like C corporations, discussed later, S corporations must file articles of incorporation, hold director and shareholder meetings, file annual corporation tax returns, keep corporate minutes, and vote on corporate decisions. Most S corporations can use the more straightforward cash method of accounting whereby income is taxed when received and expenses are deductible when paid. Unlike C corporations, S corporations are limited in the number of shareholders they can have.

Limited Liability Company

The limited liability company (LLC) is a separate legal entity established by filing articles of LLC formulation or similar documents in the state where it is formed. The number of LLC members, various classes of stock, and tax accounting selection determine a diversity of avenues to properly meet tax liabilities, whether the LLC is treated as a partnership or a C or S corporation.

Tool Shed:

Federal Employer Identification Number (FEIN)

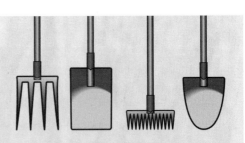

If you set up a corporation, LLC, or business partnership or find yourself hiring employees, you'll need a federal employer identification number (FEIN) from the US Internal Revenue Service. You can apply for this number on the Internet (irs.gov) or by filing the IRS form SS-4. The FEIN is sometimes referred to as an employer tax ID (EIN) or a tax identification number (TIN); they're the same thing.

C Corporation

The most expensive and complex business structure, the C corporation is a legal entity set up within a given state and owned by shareholders of its issued stock. It's unlikely, due to the scale of your operation, that you'd form a C corporation, the more common structure for companies like Tyson, Kraft Foods, and Monsanto.

The corporation, not the shareholders or directors, is responsible for the debt and liabilities of the C corporation. C corporations must file articles of incorporation, hold director and shareholder meetings, file annual corporation tax returns, keep corporate minutes, and vote on corporate decisions. Income from C corporations, after expenses have been deducted, is taxed both at the corporate level and at the individual level, on wages and dividends paid to shareholders.

Regardless of the structure of your business, you will need to file some form of annual state and federal tax returns, using either the FEIN for your business or your social security number if you're a sole proprietor.

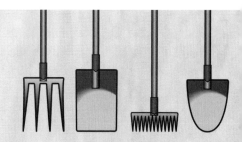

Tool Shed:

Affordable Legal Documents

Some states have very straightforward forms that must be completed to set up small businesses as a corporation or LLC. Others, less so. Below are a couple of low-cost options to consider if you want to structure your business as either. Of course, if you tend to be risk adverse and have the funds, you may feel most comfortable hiring an attorney licensed in your state who specializes in small business legal issues.

Nolo Press
nolo.com
This is one of the Internet's leading websites offering free and easy-to-understand legal information and various do-it-yourself products.

LegalZoom
legalzoom.com
For as little as $99, plus state fees, this online company can help you set up your enterprise as a LLC or subchapter S corporation in most states. It could be a source for other legal documents, too.

Depending on your comfort level, you may be best off leaving the details and specifics of setting up your business to a certified public accountant (CPA) or a business attorney. While all business structures require governmental record-keeping and forms to be filed, corporations and limited liability companies (LLCs) involve additional legal and accounting requirements. That said, many farmers I've met over the past 20 years have found the process straightforward enough that they did it themselves.

Managing health, liability, and production risk with insurance

Life is risky business, and a farming livelihood is among the more risky, according to Bureau of Labor Statistics. There are long hours, literal ladders to climb, and machinery to operate, and when it comes to animals, well, you can never be completely certain of their behavior.

From a liability perspective, producing and distributing food products, even directly to your customers at a farmers' market or CSA share box, needs to be managed in a way that doesn't keep you awake at night. If you add a farm stay or cottage food business to your diversified portfolio, you add another human element to the mix. Besides how you have your farm set up, you'll need to cover yourself should someone slip and fall on your premises.

In our litigious society, insurance has become a necessary part of both living and doing business. You need to evaluate your appetite for risk and the insurance requirements related to your business. Start by checking whether you are already covered in some way, either through existing homeowner's insurance or in how you structure your business (covered earlier). Consider combining insurance with other strategies.

If you're already in business, you're probably covered at some level for farm or personal liability. In most cases, it's nothing more than checking with your insurance company. For example, for our small farm, B&B, and cottage food enterprise, doing business as Inn Serendipity, we have the business specifically listed on our homeowner's insurance policy as a rider; the annual premium is less than $100 and provides $300,000 of personal liability coverage. Our business is viewed as "incidental to the home" by our insurer, Cincinnati Insurance Company. Additionally, we have a $1 million umbrella insurance policy that covers anything we do on our property and in our car, above and beyond the $300,000 liability coverage.

Everything changes should we decide to prepare food products in a rented commercial kitchen; at that point, we'd need to purchase a commercial business liability policy that may run, at minimum, over $500 per year. Commercial business insurance might also cover us in situations where we are doing off-farm sales of our products, like at a farmers' market or holiday bazaar. The annual premium for your business and product liability coverage will depend on the level of coverage, sales venue, gross sales volume, and, of course, your products themselves. In many cases, a commercial insurance policy will be required not by the state, but by the venues where you sell your products. Many sales venues require a "certificate of insurance" for an amount ranging from $300,000 to $1 million; this certificate, generated by your agent based on your policy, explicitly refers to the venue by name.

Writing a business plan

Before starting a farm, it's wise to get your ideas, goals, and aspirations written down. But know that such a labor-intensive undertaking is unlikely to make your farm more profitable when you're just starting out. It's the process of writing the business plan that's most valuable. It forces you to evaluate every aspect of your enterprise and get these details down on paper.

In *Homemade for Sale*, I called a simplified version of a business plan the "back-of-a-napkin plan" that covers the basics: products, objective, niche markets, target markets, positioning statement, sales venues, start-up expenses, fixed expenses and variable expenses. In case these terms might be new to you, here are quick definitions:

Products: Fresh fruits, vegetables, eggs, or pastured meat products that might have been processed in a USDA-licensed processing facility that you sell frozen

Idea Seed:

"New and experienced business owners, regardless of history or current situation, can benefit from business planning. As an experienced producer, you may develop a business plan to map out a transition from conventional to organic production management; expand your operation; incorporate more family members or partners into your business; transfer or sell the business; add value to your existing operation through product processing, direct sales, or cooperative marketing. It's never too late to begin planning! If you are a first-time rural land owner or beginning farmer who may be considering establishing a bed and breakfast or community-supported agriculture (CSA) enterprise, business planning can help you identify management tasks and financing options that are compatible with your long-term personal, environmental, economic, and community values."

— *Building a Sustainable Business*, Minnesota Institute for Sustainable Agriculture, University of Minnesota

Objective: What you want accomplish with your farm business and how you want to achieve it

Niche market: A defined segment of a larger market you've identified as a potentially financially rewarding opportunity. What makes your products different — and better? Are your ingredients grown organically or your animals raised on pasture? If you define your niche too narrowly, or if there aren't enough customers who want what you make at the price you're asking, then this niche doesn't have the market potential to make your business viable. Don't throw in the towel; just re-evaluate your goals and examine ways to expand your market without being everything to everyone.

Target market: The audience or potential customers you want to reach, serve, and satisfy with your products. You can define your target market by demographics such as age, geography, and income level, or by psychographics, such as attitudes, beliefs, and value systems. Restated, demographics help you understand who your customers are, while psychographics help determine why your customers buy what they do. Many women farmers featured in this book define their target markets as people who care intensely about what they eat, how it's grown or raised, and the health of the land. Additionally, they focus on building a regional food system and appeal to people who prioritize their household budget, selecting high-quality foods for their family.

Positioning statement: Marketing jargon for the combination of marketing elements that go into defining your product. Positioning can consider all aspects of marketing (described in detail later) plus how your product might be used or might solve a particular problem. Often, positioning can involve a combination of several variables. You're conveying what your product is, how it's different from the competition, and why a customer should buy it from you.

Sales venues: Farmers' market, on-farm stand, special order/delivery, CSA, wholesale

Start-up expenses: Could include livestock; equipment; fencing; local, state, and federal licenses (for a farmstay B&B, for example)

Fixed expenses: A business telephone line, insurance, website hosting, and domain name registration

Variable expenses: Could include seeds, soil amendments, produce crates, straw bales, gas for delivery

If nothing else, the back-of-a-napkin abbreviated plan makes sure you don't forget the great ideas you've come up with to get your farm products to market. Plus, it might keep you on track. This simplified plan may help head you off from arbitrarily chasing an "amazing" new opportunity that doesn't fit your goals or objectives. Sure, a herd of dairy goats for free might sound great, but if you don't have a plan for caring for, containing, and milking them, let alone a market for their milk, you'll have a liability, not an income-producing asset. One goat owner ended up becoming an artisanal cheese maker as a result of the volume of milk her herd produced. Think of your business plan as a signpost for profits.

If the time comes when you need more money and want to expand, you'll have plenty of time, enthusiasm, and determination to pull together a business plan that would impress an MBA professor type, lure money from the deep pockets of bankers, or secure a beginning farmer capital loan from the FSA.

There are plenty of free online resources that guide you through the steps of writing a business plan, from a vision statement to pro forma income statement and balance sheet. One of the most complete, perfectly tailored to small-scale farmers and rural entrepreneurs, is *Building a Sustainable*

Idea Seed:

"Always add humor and heart to everything you do, and use the business to express your creativity. It's your venture, and you don't have to follow tradition. We all work hard. We might as well have fun."

— Sarah Calhoun, founder of Red Ants Pants, a workwear clothing company for women, White Sulphur Springs, Montana

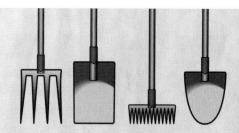

Tool Shed:

Key Resources for Your Business-planning Library

Fearless Farm Finances: Farm Financial Management Demystified, by Jody Padgham, Paul Dietmann, Dr. Craig Chase, and Chris Blanchard, published by the Midwest Organic and Sustainable Education Service (mosesorganic.org/fearless-farm-finances). This one-of-a-kind resource is packed with instructions, tips, and tools for setting up and managing a farm's financial system.

Business: A Guide to Developing a Business Plan for Farms and Rural Businesses, free from the Minnesota Institute for Sustainable Agriculture (www.misa.umn. edu) and published by Sustainable Agriculture Research and Education (SARE); it includes values, farm assessment, vision, mission and goals, and strategic planning worksheets. A few other b-plan templates that you can customize for your needs include enloop.com, score.org, and sba.gov. (Follow the links to writing a business plan.)

Besides providing resources for small businesses, the association for America's Small Business Development Centers (asbdc-us.org) offers a small-business assistance network in the United States and its territories. The network's mission is to help new entrepreneurs realize their dreams of business ownership. Small-business owners and aspiring entrepreneurs can go to one of approximately a thousand local SBDCs for free, face-to-face business consulting and at-cost training on writing business plans, marketing, or regulatory compliance.

Sell it: marketing

Because of the growth in the number of women farmers, you are the story the media often want to cover. You'll just need to sell your story on your website, Facebook page, and through those alluring visuals shared on Instagram. Marketing is a term that covers a wide range of considerations associated with selling.

Most business schools cover the four Ps of marketing: Product, Price, Place (how you distribute your item), and Promotion, both in the form of paid advertising and "free" public relations. We add three additional Ps in our *ECOpreneuring* book, in part because marketing has become pervasive and integrated into day-to-day routines, often in subtle or clever ways. It permeates our life through product placement in movies, naming rights for stadiums, Facebook updates, or reviews on Yelp. These additional three Ps of marketing — People, Partnerships, and Purpose — reflect our values and belief systems and connect us to our community. The most effective marketing efforts are those that combine all seven elements into one cohesive, integrated, and clear plan that can be effectively implemented.

Below is a brief summary of the seven Ps that should carefully incorporate your niche, target market, and positioning statement. Since the Internet and social media have emerged so strongly in many farmers' approaches to marketing, both are covered in the "Tech it" section that follows.

Product

From heirloom tomatoes to heritage breed beef cattle, from a farm-stay experience to jars of strawberry jam, your products are both the physical properties and the marketing behind them. This includes your product name, company name, logo, and slogan.

Price

Simply the price you establish for the products you sell.

Place (distribution)

Where and how you sell your product is referred to as "place." Whether you distribute CSA share boxes of fresh fruits, vegetables, and herbs or wholesale to a regional food hub, carefully evaluate the venues that ensure your profitability. While a farmers' market four hours away in a big city might be highly lucrative, you'll need to evaluate in terms of the time and physical toll it may take during the growing season.

Promotion

Promotion is what most of us think about as marketing, the applied art form of persuasive communication by graphics and words to help sell products. Creative, innovative thinking drives your approach to promotion — telling the story about your product — that leads to customers wanting to buy your products. Promotion helps them realize they have a need that will be satisfied by purchasing your product. Included in promotion is both advertising and public relations (PR). Advertising is purchased, whereas PR, whether solicited or unsolicited, is free. Depending on your skill set, comfort with a computer, time, and budget, there are many free or nearly free promotional opportunities. Despite the saying, "You have to spend money to make money," we've found that the less you have to spend in promotion, the more you earn selling your products. The trick is to find the most cost-effective way to reach your target market.

People

One way or another, your product will satisfy some need of your customers. When you're defining your target market, you may notice that sometimes the person buying a product is not the person eating it. Decide whether you are going to target the customers with the cash or the consumers who actually end up savoring the product. Your marketing can be designed to reach one or both of these audiences, albeit with different messages and goals. You're also selling community and a connection to the land. Next time you visit a farmers' market, notice the relationships and bonds between farmers and their customers. Many who frequent farmers' markets are there for the friendship, the ecological connections to the land, and a sense of hope and optimism for the future — far more than for the potatoes, pea pods, or homemade salsa. What you're selling is adding to this richness of community.

Idea Seed:

"We have only one life, and growing nourishing food is one of the most important things you can do with your time here. Teach your neighbors, your friends, and your children how to grow and care for the land."

— Diana Rodgers, founder of SustainableDish.com, a blog dedicated to optimal foods for humans and the environment, and author of *The Homegrown Paleo Cookbook*, Clark Organic Farm, Carlisle, Massachusetts

Idea Seed:

"I sold all my Shetland sheep, a breed specifically for producing wool. It's important to go with your passion in running your farm business, and I realized wool and fiber are not mine. My heart lies in food production, so all the animals I raise produce something edible."

— Deborah Niemann, author of *Homegrown and Handmade*, *Ecothrifty* and *Raising Goats Naturally*, Antiquity Oaks, Cornell, Illinois

Partnerships

While steering away from traditional paid advertising outlets, you may discover strategic partnerships that open new doors to connect with your target audience. When dollars are exchanged, such as when your business makes a donation or takes out a membership with a nonprofit organization, the money goes directly to furthering your shared mission. Explore ways you can thrive on connections with like-minded organizations. For example, can your bakery products be an add-on to a share for a farm that follows a community supported agriculture (CSA) model? By joining with nonprofit organizations or various causes to help advertise a product, your small enterprise participates in what is commonly called "cause-related marketing." You cultivate relationships that echo your values, reinforce your business's commitment to the issues you care about, and connect with people who share your interests, passions, and sense of purpose.

Purpose

Purpose-based marketing provides the ultimate in competitive advantage for small businesses. Your sense of purpose and passion can be reflected in everything you do. This compelling message, a part of your inspiring story, puts a face to the food items that your customers then savor, one bite at a time.

Tech it: Internet, social media, and the Internet of Things (IoT)

The pervasiveness of the Internet has made having a presence there necessary for any business. You'll need to decide what approach you take, which might include a website, online storefront, social media, or a combination of these. The great news is that in most cases, you won't need to write one line of code. If you can "drag and drop," you can launch a website within four hours.

There are two ways to approach website development, one that involves money and another that is free. Gone are the days when you needed big bucks for a website. Since many small-farm businesses may be just getting off the ground, starting with a free website might be the simplest and wisest choice.

1. Free websites

The following companies offer the ability to modify easy-to-use templates and customize them for your business; there are many other options as well. If you have some computer experience, the intuitive nature of the websites makes them easy to navigate, and instructional videos will guide you through the design, so there's no programming or "coding" involved.

The websites do have some space and creative limitations and may come with small ads that also appear on your website. But for most first-timers, you'll be amazed by the results. Just register for the website template you like the best and start uploading text and photos. There's plenty of free storage space.

wordpress.com

It's so easy an 11-year-old can do it. Really. This is the leading blogging interface that can be adapted easily as a business website. For the record, "blog" stands for a "web log." If you love writing about your products, ingredients, or journey as a food entrepreneur, this option will be particularly attractive.

wix.com

Containing numerous templates, many product oriented, this online website builder focuses on easy drag-and-drop design elements. Stick to their HTML5 options, since more and more people are viewing websites via their mobile smartphones, so you want to make sure your website looks okay on these tiny devices, too.

weebly.com

This is a very basic, visuals-driven website design interface. Not many bells and whistles, but its simplicity will appeal to less tech-savvy people and get you quickly set up on the Internet.

sites.google.com

Created by the most widely used Internet search engine company, Google Sites provides the ability to create a free website with various features. If you like to write regularly, then you can use Google's blogspot.com.

2. Low-cost websites

More experienced entrepreneurs who want greater control over their name, products, design elements, and website capabilities can purchase their domain name (the Internet name you select to represent your company) and then host their own website. Both the domain name and hosting fees cost less than $100 per year from companies like GoDaddy.com.

Hosting and designing your own website may require greater computer knowledge than you have the time or interest for. If so, you could hire a website designer, depending on your budget, your design goals, and the scope of what you want the website to do for your business. Some farmers find a family member, friend, or neighbor who is happy to help design a professional-looking website,

Idea Seed:

"Being yourself is the best marketing there is, and the blog gives me a space to be myself in an interactive way that our website does not. The general population may not be ready to commit to a farm or even be interested in agriculture, but they all still need to eat, and everyone is always looking for new recipes. I share personal, honest tidbits on the blog about my relationship with my husband, Kyle, or the difficulty of this path we're on or what have you, just life in general. I'm very open about our life and our choices. People want to know you, people want to understand you, people want to relate to you and believe in and support you."

— Lauren Wells, Raleigh's Hillside Farm, Brodhead, Wisconsin

perhaps in exchange for a regular box of fresh produce throughout the growing season.

Social media

Everything you do is about sharing your story. Don't overlook ways to let your customers do this for you as well. To build awareness around your product, you'll need to get people to try it, love it, and share what they like about it with the rest of the world. Social media have become an increasingly important part of an advertising campaign. And it need not cost you a penny to get started.

Lauren Wells with watermelon, Raleigh's Hillside Farm, Brodhead, Wisconsin.
KYLE RUDERSDORF

Your most effective advertising is the satisfied customers themselves. Word of mouth has always trumped a four-color display ad in a magazine. People are much more likely to trust their friends than a company trying to sell them something (even if what you're selling is great).

Thanks to the proliferation of social media, there are lots of options for sharing your story with the world, in characters, updates, photos, and video. The multiplier effect cannot be overstated. But it does require a different modus operandi, where talking becomes typing and a printed poster becomes a "folder" of incredible photos of your products, your farm, and your customers relishing a bite of what you've produced. Because there aren't enough hours in the day to do them all, carefully select the social media your customers use most.

The more your customers rave about your products to others on the Internet, the better. People who love your product can, in spirit, be your "in-house" advertising agency. They can tell their friends, share links to your products on Facebook, and tweet about their favorites, too.

Facebook: facebook.com

Currently, this is the dominant social media networking service, where you can keep connected to your customers and share regular updates, such as a new product or an event you'll be selling at next week. As part of marketing strategy, some companies are now choosing to make their business Facebook page their de facto "website."

When you start your venture's Facebook page, be sure to select and create a "business" profile, not a "personal" page. This keeps your business professional and opens up opportunities you won't have on your personal page, including the ability to schedule posts in advance, assign other people as administrators (to help you), access analytic tools, and implement targeted advertising campaigns, if you choose to do so down the line.

Google+: google.com/+/business

This search engine giant likewise has ambitions to thrive in the social media world through Google-plus for business.

Twitter: twitter.com

If you like texting, then this online microblogging website is perfect for sharing what's happening with your business in 140 characters or less.

Pinterest: pinterest.com

Think cork bulletin board with photos, embedded on an Internet page. This pinboard-style website can spread images of your food products through the Internet if your photos are beautiful enough.

Instagram: instagram.com

This is like Twitter, except that what you share is a snapshot, not text. So when your harvest that first tomato, share that candid photo here. Instagram currently works only with mobile devices.

YouTube: youtube.com

A free, movie-sharing website, particularly useful if you have the talent and interest in creating videos around your products and their use in the kitchen, perhaps as cooking demos.

LinkedIn: linkedin.com

This professional business network can connect you to people once

Idea Seed:

"Instagram and Facebook enable us to continue to stay in touch with our customers between markets or early in the season before we start CSA pick-ups but when things are happening on the farm."

— Jen Miller, Sandhill Family Farms, Grayslake, Illinois

impossible to reach. It's all about degrees — or links — of separation. Try it out and see if you're only six degrees of separation away from everyone you might want to meet.

Just because you have 742 Facebook "friends" doesn't mean these friends see every "status update" you post. Make no mistake, not only are Facebook and other social media sites mining personal information about you and your online life, they're making money off you, too. In fact, if you have "cookies" disabled on your browser, you cannot even sign in to use the site; cookies track everything you do. Facebook and many other social media companies have proprietary and secret algorithms they use to control how many people see your updates.

If you want to boost your reach and increase your audience on social media, you have to pay for it. This is called "pay-for-clicks." You can focus on people who like your page and their friends, or broaden your reach to people you target. Right on their main business page, the social media site will show you how you can increase your reach and how much it will cost you; heck, they even create a sample advertisement out of the content you just provided.

All you have to do is enter your credit card and set your parameters, including your budget, target market, and duration of the campaign. Then with a click of a button, your ad will reach a segment of the population so specific that it's a bit creepy, at least to us. The good news, however, is you can effectively target a market at a potentially very low cost; your update will show up in their "news feed."

The Internet of Things

Imagine getting a message on your smartphone or laptop, notifying you that field three with your summer squash and peppers needs to be irrigated. When that goat gets out, again, you will not only be notified that it escaped, but you can track it with your smart watch. Any human-made or natural object can be assigned an IP-address just like you have on your laptop or desktop computer.

According to Tech Target, "The Internet of Things (IoT) is a scenario in which objects, animals or people are provided with unique identifiers and the ability to transfer data over a network without requiring human-to-human or human-to-computer interaction. IoT has evolved from the convergence of wireless technologies, micro-electromechanical systems (MEMS) and the Internet."

Kind of like *Star Wars* meets *Little House on the Prairie*, such technology and the farm form interesting dance partners. On one level for many of us, technology enables us to exist, from our website to the sunrise Instagram we send from the field to our CSA members. But we also personally need to feel that soil running through our fingers and to know our ewe has mastitis just by looking at her, not because our smartphone started beeping. Balance being open to new advancements with continuing to grow that encyclopedia in your mind.

Do it: getting down to business

To be legitimately in business, you need to have made at least some profit in at least three of the past five years, according to the IRS. Business revenue minus expenses equals profit.

By being in business, and not just having a hobby, you can deduct expenses related to operating your farm. IRS-recognized expenses include the cost of seeds, livestock and feed, soil amendments and fertilizer, wages (including paying yourself), interest payments for farm-related loans, depreciation related to farm equipment, and insurance premiums for your farm.

In a year when expenses exceed revenues, the business loss can be claimed on your personal tax return and may end up reducing your income tax liability if you have other sources of income. For more clarification on this, consult with a CPA or other tax professional.

The process of tracking the money coming in and going out of the business is called bookkeeping. You can keep track of sales and expenses with a ledger, either electronically or on paper. You can also use your ledger to record transactions related to assets, liabilities, and owners' equity. There are numerous tablet-based bookkeeping apps. For the more computer-savvy crowd, Intuit's popular QuickBooks (quickbooks.com) may suffice, but you may find these too detailed and expensive for your needs, at least when you're just starting out.

To keep records, you can put sales receipts in one folder, box, or cabinet drawer for each month, quarter, or year, depending on the scale of your operation. If you don't have a receipt system, create a simple receipt on paper with the date, customer/event, and total sales. For expenses, hang onto every receipt, from your start-up costs to ingredient purchases, again, organized by month, quarter, or year. These receipts are proof that you're in business. You don't have to make it any harder than this.

Two methods can be used for accounting, the process by which financial information is recorded, summarized, and interpreted. The more commonly used and straightforward "cash basis" of accounting establishes that income is taxable when received and expenses are deductible when paid. The more complex "accrual basis" method records payments or expenses when they're agreed upon (e.g., the date an invoice is issued), even though the actual transaction may take place at some future date (e.g., the date the payment is received). Most farmers opt for cash-basis accounting due to its simplicity. Go out of your way to deposit every last cent of revenue you earn, and document your fair and reasonable business expenses.

A business bank account is a must. Co-mingling personal and business assets will invite an audit by the IRS. Another red flag is not being "fair and reasonable" in terms of your expenses. Going to a farming conference or deducting a pick-up truck used exclusively for business is both fair and reasonable. Taking the family to Disneyland, not so much.

Forms of payment

You'll also need to decide whether you want to accept checks or credit cards in addition to cash. Following are some credit card-processing options that charge you a nominal fee based on a percentage of the transaction. Thanks to the proliferation of mobile devices and computers, processing credit cards has become easier and more widespread. Most companies offering "card readers," small devices that can read a swiped credit card, also provide an option to manually key in the credit card number, but they charge a higher percentage fee and fixed transaction cost for this feature.

PayPal: paypal.com

Among the most widely used, secure, and safe ways to receive payment via credit cards is through PayPal. No contracts or monthly service fees; just a flat fee plus percentage for each transaction. PayPal offers a mobile app and card reader for payments on the go.

Square: squareup.com

Using a free Square device that plugs into the phone jack on your smartphone, tablet, or computer, you can swipe the card and complete your check out from just about anywhere.

Spark Pay: sparkpay.com

The free card reader from Capital One Bank can process credit cards on a smartphone, tablet, or computer using wireless and mobile devices.

Money belt or drawer

You'll discover that few people pay with the exact change. Depending on the sales venue, customers may arrive fresh from a cash station with twenties or fifties. To cover your cash sales, you'll need to have a system for making change. A money belt or cash drawer, besides being convenient, helps avoid bills and change being placed in your pockets or coat jackets in the flurry of business transactions.

Cash flow is queen

Many businesses get into trouble, not because they don't have great products, but because they don't manage cash flow well. Cash flow is simply the cash received and spent by your business over a specific period of time, usually a year. Farming is a particularly risky enterprise when it comes to managing unpredictable — and increasingly severe — weather and other natural calamities. Add to this a highly seasonal nature and, at times, unpredictable or fickle customers. A brief heavy thunderstorm just prior to the opening of a farmers' market can result in dismal sales that day, even if your lettuce mix couldn't be better; what to do with those bags of highly perishable salad greens further complicates matters. Plan for such situations to the extent you can. Always set aside as much revenue as possible for a rainy day.

For planning purposes, a cash flow projection, on a monthly or quarterly basis, includes both anticipated cash receipts from sales and cash disbursements for expenses, including cost of goods sold (direct costs related to your products), marketing, license fees, delivery costs, and equipment. The goal is to avoid or minimize any periods where you have a negative net cash balance.

As in life, your farm may have ups and downs, crop failure, and cranky customers. However, if you follow the guidance offered throughout this book, you'll hopefully sidestep most of the trapdoors and find enough of the secret passageways so that your journey is fun, satisfying, meaningful, and still financially solid and rewarding.

Chapter 8

Your Body

Y OUR MOST IMPORTANT TOOL on the farm isn't your Farmall, Stihl, or Troy-Bilt. You won't find it stored on the barn shelf or in the mechanic shop for repair. In theory, it comes with a lifetime warrantee, as it will be with you every day you're on this Earth. It's you. Your body. All 700 skeletal muscles, 206 bones, and one and a half gallons of blood make up the most powerful, versatile piece of equipment you'll ever have.

Just as you change the oil of the tractor on schedule, sharpen your chainsaw blades before stocking up on wood for the winter, or drain the gas out of the rototiller at the end of the season, care for that amazing, multipurpose body of yours. By prioritizing our bodies, we operate better as farmers. In a society where we increasingly sit on our rears in front of screens, an agricultural lifestyle gifts us with daily routines that keep our bodies healthy. By operating with thoughtful and proactive care, we lengthen the life span of our agricultural livelihoods.

In this chapter, we explore how to stay vibrant and healthy throughout our life span. By making conscious decisions that both feel good and care for our bodies, we stack the odds in our favor to farm long and prosper. We'll first look at physical body care and then dig deeper into tool selection and clothing choices.

Live it: body care

Whatever our age, let's embrace and celebrate the reality that we will be a day older tomorrow. Especially a priority for those of us encore farmers embracing

agriculture mid-life, prioritizing preventative care — what can we do today to keep tomorrow healthy — takes on a sense of urgency as we continually lose overall strength as we age.

Ann Adams and Liz Brensinger champion the importance of women running small-scale, mostly non-mechanized farms to prioritize physical care. As co-founders of Green Heron Tools, an agricultural entrepreneurial venture that creates ergonomically correct tools for women farmers, their passion for supporting women farmers to care properly for that most important tool — our bodies — runs deep. (See "How She Sows" it for their full story.)

Adams and Brensinger offer four tips for practicing healthy body care in daily farm routines that will add up to long-term physically strong livelihoods in agriculture:

1. Plan your day

Plot your farm tasks ideally so you can scatter a diversity of tasks throughout the day. "Varying your activities and not doing any one thing for too long is one of the best strategies you can do for your body," advises Brensinger. "Personally, now that I'm in my fifties, I'm more acutely aware of my body after I've been working in the same position for too long and know to change tasks."

Break up longer chores into smaller segments: If you are moving a large compost pile to your vegetable rows, after every couple of wheelbarrow loads, change motions and squat and find a row to weed for a while.

2. Warm up

Just as you would before engaging in a sporting activity or workout at the gym, warm up your body before starting strenuous tasks. Walking is a great warm-up activity for anything you do. For example, walking to your outbuildings is a good time to stretch and get your muscles moving before doing something more strength focused, such as lifting feedbags.

Starting and ending each day with basic stretches also helps get your body ready for daily farm action. (See "Tool Shed" for some ideas.) "Before you stretch, remember to first get your blood circulating and heart rate up a bit by walking in place," offers Adams, a former nurse.

Another "warm up" technique is for your brain: before you do anything else, write a list of what you need to do that day and when. Even if the day doesn't go exactly as you plan, the act of writing that list clears your mind so you can better focus on the task at hand.

Debra Sloane, Sloane Farm, Washington, Connecticut

Let's Get Physical for Farming Long Term

"I want to be as much as possible a one-person operation. Therefore, I need to be physically strong, from my knee joints to my shoulders and everything in between," shares Debra Sloane, a women in her 50s who left the high-tech corporate world in California and moved back to her New England roots to start an organic market garden farm, Sloane Farm. She produces a lunch salad for a local shop, grows vegetables for a 10-member CSA, and grows strawberries for a pastry chef and for another woman farmer who makes chocolates. "After I finished that first growing season two years ago, I knew first-hand that this is hard work. I quickly realized that the better shape I'm in, the better I can handle everything."

A lone ranger farmer, Sloane radiates commitment to caring for her physical health. She knew there was risk involved with starting a farm mid-life, and folks along the way doubted she could do it. "That's when my stubbornness kicked in, as I wanted to prove the naysayers wrong," Sloane adds. "A big part of my success was my being able to handle the physical work." Sloane supplemented her income during the start-up years with about 30 hours a week of consulting work, wearing her old corporate hat but working from the farm.

Sloane's big piece of advice is training year-round and balancing physical activities on the farm with training at the gym. Sloane discovered she operates best when she dedicates an hour at her local CrossFit four days a week and runs the other three days. "I first thought going to the gym would be only a winter thing, but little did I know I'd not only become hooked on this high-energy workout, but I'd also develop significant core and upper body strength," she explains. Getting off the farm and connecting with a supportive workout community also helps; she likes working out between 6 and 7 AM so that she has her whole farming day ahead of her.

Know your body and use the winter months to address chronic issues. "At the end of one summer, my lower back hurt after picking bush beans," recalls Sloane.

"During the following winter, I prioritized working on my back by doing yoga stretches like child's pose and superman, and also trained my core doing sit-ups with weights." That preventative strategy proved effective, as her back was pain-free the following season, right through the bean harvest.

"The growing paleo movement turned out to be a real niche marketing advantage for a small-scale grower like myself," shares Sloane. "Every vegetable in my box is considered part of a paleo diet, which means I don't grow white potatoes, corn, or peas." A bonus for Sloane: The health-conscious friends she made at her gym add up to an easy drop-off site for her paleo CSA.

Sloane's advice, particularly for solo farmers starting an encore farm: You have to love it. "Farming has to be something that is your source of pride, the drive that keeps you going through the heat of summer when you get so hot and sweaty that the bugs stick to you," says Sloane. "It's a labor of love."

3. Know your body

Take the time to understand your body and be able to recognize what your body needs and what works well for you. Alissa Moore, a Fredonia, Wisconsin, farmer in her 30s, felt some back pain while simply sitting and didn't ignore the issue. "I need my body to be in good shape because I want to keep doing this into my fifties and sixties and beyond," explains Moore, who runs Wild Ridge Farm, a diversified vegetable operation.

She connected with a physical therapist who helped her look closely at various aspects of raising vegetables and gave practical suggestions on how to make those tasks more ergonomic. For example, when reaching down to pick something up, Moore is much more conscious of bending her knees versus keeping her legs straight and bending at the hip.

Vary your tasks and movements, and don't do one thing for too long. Many short-term as well as chronic aches and pains are a result of too much repetition of the same movements or postures. Whenever possible, stand for a while, sit for a while, and then stand again. When standing, periodically shift your weight from side to side and do some knee bends. When sitting, it's best to have your legs uncrossed to facilitate optimal blood flow and avoid putting pressure on any particular area. Break up your static motion with short breaks. There are always short chores to be done that offer the perfect break: get the mail, check the livestock water, and make sure the kids eat lunch.

4. Be conscious

Be conscious and aware of movement. "People fall at any age, but the consequences of our falls increase tremendously as we age," explains Adams. "The main way to avoid falling is to be present in your body and conscious of your surroundings and what you are doing."

Don't try to do too much at once or be preoccupied thinking about your next task. Fully engage in and enjoy your task at hand and complete it successfully and safely before moving onto something else. If it's a big task, break it up and intersperse other activities for diversity.

With some advance thought and priority on prevention, farming can be a wonderful way to stay fit and healthy and give the world a smile when you wake up in the morning. No matter how long you march on the treadmill at a gym, it will never gift you with carrots, cabbage, or cucumbers!

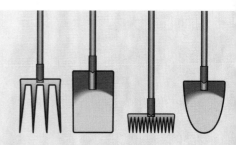

Tool Shed:

Daily Stretches for Good Health

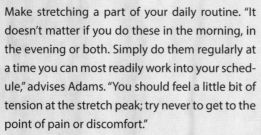

Make stretching a part of your daily routine. "It doesn't matter if you do these in the morning, in the evening or both. Simply do them regularly at a time you can most readily work into your schedule," advises Adams. "You should feel a little bit of tension at the stretch peak; try never to get to the point of pain or discomfort."

Please note that if you have an injury, disability, or chronic pain, it's important to consult a health care professional before attempting any specific stretches or strengthening exercises.

The Green Heron website provides clear explanations and illustrations of stretches and offers a wealth of other ergonomic resources.

www.greenherontools.com

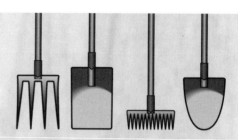

Tool Shed:

Lift and Shovel with Love and Care

Green Heron Tools shares proper technique for lifting and shoveling:

Lifting:

- Use your much stronger leg muscles for lifting. Not your back!
- Bend at the knees and hips, keeping your back straight.
- Lift straight upward in one smooth motion.
- Ideally, women should not lift more than 30 pounds. Men should keep it at not more than 50 pounds.

Shoveling:

- Stand with feet a comfortable distance apart and keep back straight. Knees should be "soft," meaning slightly bent and not locked or rigid. Keep elbows close to body.
- Use leg muscles to push the blade into ground. Keep wrists in a neutral, unbent, position.
- After the blade is inserted, slide your non-dominant hand about halfway down the shaft of the shovel while bending at your knees, not your waist.
- To lift the soil or other material, straighten your knees back into that soft and slightly bent position. Lift with your legs!
- Throw material straight ahead or pivot your body in the direction in which you want to throw the material.

Ann Adams of Green Heron Tools demonstrates proper lifting technique. Green Heron Tools

Work it: tools for women

Think of any tool, whether it's a simple shovel or mechanized chainsaw, as an extension of your body, something that should work in tandem with the rest of you. But the female body is uniquely different from that of those boys. Yes, I know you knew that part already! But here are some specifics that may be new to you. As women we have

- significantly less upper body strength; our strength is in our lower bodies
- narrower shoulders and wider hips
- proportionally shorter legs and arms
- greater flexibility
- less height than the average male — five inches on average

The trick is to recognize these physical differences and account for them, particularly as we age. "Changes in muscle mass are a completely normal part of the aging process, but it is important to be aware that our bodies simply do not have the same level of strength we used to have and to find alternative ways to accomplish tasks," advises Ann Adams. For example, even though you may have been able to lift that 50-pound bag of feed single-handedly in the past, as you get older, divide that feed bag into two 25-pound loads or use tools to help you lift and move, such as a wheelbarrow.

Idea Seed:

"Now that I'm in my sixties, I practice yoga every morning and attend yoga class one night a week in town. The stretching movement helps support my joint health and the whole experience of getting off farm and going somewhere else where I only need to follow the teacher's instructions relaxes my mind. I barter vegetables with my yoga instructor, so it's a win-win for both of us. We've become good friends."

— Denise O'Brien, Founder of the Women, Food and Agriculture Network; Rolling Acres Farm, Atlantic, Iowa

Our joints are looser than men's depending upon where we are in our menstrual cycle, which may make us more vulnerable to injury, from exposure to vibration. Bone density also naturally breaks down as we age, so it's important to consistently exercise to keep our bones strong. Fortunately, farming gifts you with the types of exercise you need to keep bones strong, such as reasonable lifting. "Anything that puts stress on the bone will keep it strong," explains Adams.

Idea Seed:

"Hydraulics are your best friend. With the touch of a lever you have the same upper body strength as a man. Don't stress your body if you can make the tractor, a winch, or come-along do the work for you. Women are good at using their heads and getting everything set up for success."

— Kate Stout, owner of North Creek Community Farm,
Prairie Farm, Wisconsin

When selecting tools, look at your own health history, identify challenging areas, and make adjustments accordingly. If you know you have problem knees, for example, focus on building and using your back muscles more. Good tool design elements allow your joints to stay in a non-twisted or neutral position with handgrips that are not too constricting or so big you have to stretch.

"Tools should be ergonomic, which means designed for maximum comfort, efficiency, safety, and ease of use in the workplace," explains Brensinger. "At Green Heron Tools, we developed the concept of 'hergonomic' to describe tools and equipment designed to be safest, most comfortable and more effective for women."

Green Heron Tools' focus groups and research found the number one tool women asked for was a shovel. So that was their first product: the HERgonomic Shovel-spade (HERS) designed specifically for women. HERS is based on women's shoveling styles: the blade has an enlarged step to make the most of women's lower body strength. It also comes in three shaft sizes to better accommodate varying women's heights.

When selecting tools, strive to find tools that are

- adjustable and can be changed to accommodate users of different heights. A wheel hoe, for example, can be easily customized for you.

- allow your back to stay straight, and minimize bending. Using long-handled versions of tools like the CobraHead weeder or adding handles to shovels and rakes also provide lifting leverage.

"I can get a little geeky when it comes to my tools," confesses Clare Hintz with a grin. "But I've learned that ... the right tool investment can make the difference in both the physical and fiscal health of my farm." Hintz runs Elsewhere Farm, a diversified market garden farm championing permaculture design principles in far northern Wisconsin near the shores of Lake Superior.

Hintz evaluates tool purchases based on five criteria:

1. **Will it pay back within one year?**
   "I need to keep my bottom line first and foremost in mind," advises Hintz. "My strategy is to take a prudent fiscal approach and avoid debt as much as possible, so I'm always evaluating purchases based on payback."

2. **Is it something I can share?**
   "If it's something I only need intermittently or seasonally, can I purchase it cooperatively with other local farmers?" Hintz successfully used this strategy to buy a walk-behind tractor together with three other farmers, as it's something she only uses a few times in the growing season.

3. **Is it something better to hire out?**
   "If I'm contemplating something bigger and more expensive, I ask if it would be financially wiser to hire someone to do that work," Hintz adds. "It's a fine balance between being undercapitalized versus overcapitalized, but I also don't want to get too ahead of myself when it comes to owning equipment or any major project on the farm because my vision constantly evolves. For example, the packing shed I need today is different than the one I'll need twenty years from now, so I don't want to go overboard today."

4. **Can it change and be flexible as I grow? Can I resell it?**
   If something can be modified to evolve as the farm evolves — or if it's equipment that there's a secondary market for, like a trailer — that's always a plus.

5. **Is it saving my body?**
   "I'm always conscious of tools I can grip well and that do not require me to bend over and instead allow me to keep my back straight," adds Hintz.

What's Hintz's top tool investment for the all-around farm workhorse for women? "Hands down, I recommend a walk-behind tractor.... It's powerful but not over powering, even for my height of five feet two inches." With two self-propelled wheels, the walk-behind can make very narrow turns and get into small places. Hintz uses a Grillo brand, made by an Italian company in business for over 50 years.

"I love that my Grillo works with a variety of interchangeable parts that attach to the power take off so I can use it for anything from a tiller to a

Clare Hintz of Elsewhere Farm uses her hori hori knife.

cultivator to even a snowblower." Most walk-behind tractors start at around three thousand dollars, far less than a riding tractor and cheaper than buying multiple pieces of equipment. "Walk-behind tractors also typically have adjustable handles, which makes it easy to share with other farmers. I share mine with two guys who are over six feet tall. It's easy to move around as it just weighs seventy-five pounds," adds Hintz.

When it comes to hand tools, Hintz generally finds products made by Japanese manufacturers a better fit as they are designed for shorter people. She has three favorite go-tos:

1. **Serrated knife**
"I use a *hori hori*, which is the Japanese name for a heavy, serrated, multipurpose steel blade that's great for gardening jobs such as digging or cutting." The blade is sharp on both sides and comes to a semi-sharp point at the end.

2. **Clippers**
"Invest in a good-quality pair of hand clippers, something you can take apart to sharpen the blades," recommends Hintz. "Try different brands to find a hand style you prefer. I personally like Felco."

3. **Hoe**
"Most folks are probably familiar with the old-fashioned square-shaped hoe, but I prefer the newer triangle-tipped style as it offers more maneuverability between vegetable rows," sums up Hintz.

Championing proper tools is a political act. "Stop living with things that don't fit your body," adds Hintz. "We need to give feedback to the companies that make these tools and the retailers that sell them. I'm always sharing my opinions with my local hardware store, and they listen to me because I'm a loyal customer, and they stock my recommendations."

Ann Adams and Liz Brensinger, Green Heron Tools, New Tripoli, Pennsylvania

Creating Tools to Empower Women Farmers

The phrase "necessity is the mother of invention" fits the story of Green Heron Tools. While the number of new women farmers continues to grow, you wouldn't know it by the farm tools specifically designed for women in the retail aisle. Most items, from hand tools to larger equipment, are still designed for a male body that is taller and generally stronger. Enter Ann Adams and Liz Brensinger, female farmers turned entrepreneurs.

Avid gardeners for years, Adams and Brensinger started a small commercial farm operation as a new career venture in their early 50s to help supply Adams's son's restaurant in Pennsylvania. The duo quickly realized they lacked the proper tools to efficiently and safely farm at a scale they needed to be a viable, sustainable business. But whereas many of us female farmers would simply "make do" with our tool situation, these two embraced the opportunity to both launch an entrepreneurial enterprise and help champion this growing movement of

The women powering Green Heron Tools: Ann Adams and Liz Brensinger.
GREEN HERON TOOLS

women farmers by focusing on helping women farmers use their bodies properly through ergonomically appropriate tools.

"Our first step involved researching what tools already existed for women farmers," explains Adams. "We thought we could just collect what existed and perhaps put it all in one catalog."

But to their surprise, they didn't find any farm tools designed specifically for women. "We didn't find anything other than a few so-called 'ladies' tools that were pink or flowered and usually flimsy, as if all women needed were 'pretty' tools. We found that pretty insulting," Brensinger adds. "Nobody had ever considered that women would do better with tools designed specifically for our bodies.... Women's bodies work differently than men's and we work better and safer with tools designed for our physiology."

The duo launched Green Heron Tools. They had previously named their farm Green Heron Farms because green herons nested on their property — quite serendipitously, as the green heron is the one of the few birds documented to use tools such as small sticks for tasks like fishing. "Follow your intuition and everything feels meant to be," says Adams.

The duo received a Small Business Innovation Research Grant (SBIR) from the USDA, which supports research that leads to agricultural innovations. This led to a partnership with engineers from Penn State University and intensive research with women farmers, studying female ergonomics. Collaborative research forms the core of the Green Heron Tools business: listening to and surveying women growers and understanding what this group uniquely needs, then designing products accordingly.

"The cooperative spirit of women farmers brought this venture to life," adds Brensinger. "Women by nature are supportive and collaborative, and we are very indebted to the support and feedback we received."

Beyond the research and the business, however, Ann and Liz share a passion for helping women better understand and use their bodies from the start and offer two pieces of advice:

Minimize risk

"A core learning for us is that too many women take physical risks out of necessity. We simply want to get the job done as fast as possible and don't think that we should do something a different way, like using a lever as opposed to lifting something directly," shares Adams. Stepping back and thinking through a task before jumping in can go a long way in promoting safety and health and shrinking risk.

Think prevention

"Prevention is the best strategy to protect our bodies. Vary tasks and don't keep your body in any one position for too long," explains Brensinger. Prevention is important because we want farmers to do what they love for as long as they can. If you take care of yourself, you avoid injury that can not only hurt yourself but hurt your bottom line too if you can't work to get your harvest in. Unlike other professions, farmers don't get sick time or readily have someone else do their job.

Wear it: wearables that work

"Ever since the gold rush, women have been wearing men's pants when they need to get serious work done, and I felt it was high time to change that," proclaims Sarah Calhoun, owner of Red Ants Pants, the first company dedicated to making work clothes for women. Raised on a Connecticut farm and a dedicated outdoors woman, Calhoun grew frustrated, as she could never find a pair of durable, comfortable work pants that fit her female body. "I didn't want to start a business; I only wanted a pair of damn pants that fit." Red Ants Pants is based in White Sulphur Springs, a small ranching town in the middle of Montana. The pants are fully made in the USA.

Turns out women come in a lot more shapes and sizes in our lower halves than boys do. Men tend to be fairly square in shape from the waist down, whereas women have more variation, with curves. "Men tend to carry any extra weight in their bellies, but we women can carry it in a variety of places such as hips, butts, thighs, and even face and arms. We're apple-shaped, pear-shaped, and every fruit you can imagine," says Calhoun, laughing.

Calhoun created 74 variations of basically the same pair of pants, adjusting for size and whether a woman has a straight or curvy build. "Work clothes that fit properly serve a safety function, as we need a range of motion to move quickly and safely, particularly when we're operating machinery or up on a ladder," adds Calhoun. With this in mind, avoid clothing with drawstrings or baggy material. While something loose may feel comfortable, it's a safety hazard as it can get caught in equipment.

While quality workwear from a place like Red Ants Pants is an investment at $129 a pair, women farmers creatively make do with what is available

Sarah Calhoun of Red Ants Pants.

locally and in their budget. For the hot summer months, I keep fully covered in light cotton clothes from the thrift store. In this case, I'm primarily weeding or harvesting, so baggy flowing material works fine and keeps me cool. Covering my legs and arms completely both helps with sun protection as well as dirt removal at the end of the day. Way easier to shower off sweat than scrub off soil.

For that reason, when working the field rows I always wear old gym shoes and old white cotton socks — my "farm socks" I affectionately call them, as I don't worry about bleaching them super white ever again. My feet may get a little hot sometimes, but this system keeps my feet easier to clean at the end of the day than Teva sandals.

Tool Shed:

Work Clothes Built to Last

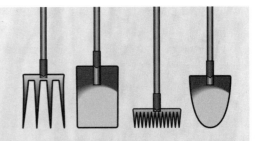

Companies that carry workwear lines for women, recommended by women of the "Permaculture Women in Practice" Facebook group:

Red Ants Pants (www.redandspants.com)

Duluth Trading Company (www.duluthtrading.com)

Carhartt (www.carhartt.com)

Arborwear for Women (www.arborwear.com)

One final wearable that women farmers use is the fanny pack. Classic 1980s retro style, a fanny pack keeps your essential gear easily accessible and works exceptionally well if you like to work in stretch yoga pants sans pockets. With my allergies, I always have tissue and my inhaler. I also carry a notepad; so many ideas pop into my head when my fingers dig into the soil that I've learned to write everything down immediately. Other secrets you'll find in female farmer fanny packs: phone, zip ties, knives, and snacks.

Idea Seed:

"When Mother Nature gives you a blizzard, it's time for the one-piece bodysuit. Somehow, your body leaks heat around the bottom of your coat, no matter how snug it is. With a bodysuit, like one from Carhartt, you stay significantly warmer. Just a law of physics, I suppose."

— MaryJane Butters, MaryJanesFarm, Moscow, Idaho

PART 4:
CULTIVATING QUALITY OF LIFE

Nurture your body, mind, and spirit

- Count your blessings.
- Always have something you're tinkering with.
- Be nice. The world is really one small town.
- Sleep outside on full moons.
- Keep calm and grow something.
- Remember that you can't solve a problem with the same mindset that created it.
- Have a hobby unrelated to farming.
- Hang hammocks.
- Recognize that loam wasn't built in a day.
- You are not alone.

Chapter 9

Sister Share

THE WOMEN FARMER POTLUCKS in my "farmhood" take a hiatus in the winter months. After years of last-minute cancellations due to winter storm warnings and icy back roads, we decided to skip our gatherings from December through early spring. The temperatures still linger in the nippy 30s, and lumps of snow inevitably dot the fields in March, but we feel the urge to get out after being cooped up during winter.

By March, we're ready to connect, share stories of birthing lambs and starting seeds. It's a celebration of getting back into the growing game. After a few glasses of wine, the late nights of lambing or assisting in goat births are forgotten. We share photos of our kids, ewes, calves, and piglets on our smartphones. It's our vernal seasonal cycle.

While the dinner buffet always showcases a range of creative culinary flare, the offerings get rather funky at this March potluck. Since we have more than 20 attendees, we always go around the room and do introductions; some of us know each other well, but there are always first-timers. Our calling card is the dish we brought, and we each give a short description.

"It's a kind of stew with barley, broccoli, cheese, and whatever else was in the fridge," shares Katharine Kramer of Katharsis Meadows, a diversified farm raising free-range chickens and Muscovy ducks for eggs and meat; goats for goat-milk soap, meat, and fleece; and pastured Berkshire pigs. We nod approvingly. We know that whatever comes in her casserole dish we want to sample. Kramer also hosts monthly themed *prix fixe* dinners showcasing her farm-to-table fare.

We taste free-range ingredients at their finest. Kramer's dish is typical of the epicurean experiments everyone brings this time of year. Mix up spring fever with the fact that our group is as dedicated as I am to using up home-grown, end-of-the-winter ingredients from the root cellar or freezer, and we know that whatever we serve has a receptive audience. The dishes, however, are one-hit-wonders. Don't ask for recipes. None exist. Simply savor and enjoy until the pot is empty.

As we do this round robin of our names and concoctions, we open the floor for anyone to mention if there's something they need. This could be needing extra tomato seedlings or looking for a local tractor mechanic recommendation because your John Deere clunked out.

Requests for information and advice always top the list. How do you make it work with a baby? How do you make a living and find a market outside of the rich elite? How do you deal with feeling intimidated at the mechanic shop when the owner assumes you don't know how to start the engine?

At our last March potluck, Peg Sheaffer of Sandhill Family Farms grew teary when she queried the crowd on how to help a sheep with the birth of triplets. She had lost both the momma sheep and the babies in that scenario only a few days before, and the grief was still raw. Sheaffer runs a successful CSA serving Chicago. In partnership with her husband, Matt, and another family she runs two farm properties serving over 500 families in the CSA plus thousands more at farmers' markets. In addition to the produce share, customers can also add a dairy, meat, or fruit share, which Sheaffer manages by partnering with other area farmers and food artisans.

We look up to Sheaffer: Willowy, strong, tall and smart, she's a power-house of seasoned farming knowledge and accomplishments in agriculture. However, everyone needs a confidant, a friend, a shoulder to lean on. Our gathering served that role as she received empathetic support along with advice for next time. Offers of "Call me and I'll come to help" echoed around the room.

Whether it's a potluck like ours,

Idea Seed:

"Your life as a farmer is very different than most of your friends' lives. You'll lose some friends because they won't understand why fixing a fence is more important than shopping, why a good night's rest before a day of chain sawing is more important than a night out. Let 'em go. You've got the richer life."

— Catherine Friend, author of *Hit By a Farm: How I Learned to Stop Worrying and Love the Barn* and *Sheepish: Two Women, Fifty Sheep, and Enough Wool to Save the Planet*

a woman you sit next to at a farming conference, or a gal you connect with online, women farmers cover topics related to our farm enterprises but also address subjects outside traditional information portals. You won't find a tip sheet at our local Extension office on how to milk a cow while carrying a baby in a sling, but undoubtedly there's a fellow female farmer out there whose been down that road and can happily help you out.

This chapter explores topics that don't fit anywhere else in the book:

- Improving communication with men
- Fitting in
- Finding your local tribe
- Integrating family and kids

Consider this chapter your virtual potluck: a private, safe, and inclusive gathering where we share our needs and find solutions. Due to the unique nature of this chapter, I've echoed our potluck gatherings. At the start of each of topic, you'll meet a fellow female farmer asking a question and opening up dialog. Through open questions and discussion come solutions and support. So, pass the cheese plate and pour yourself a glass of wine and let this party begin.

Improving communication with men

"Every time I head to the farm supply or feed store, I feel like everyone is watching me. I'm usually the only woman there — or at least I clearly am not a farm wife running an errand for her husband. When I ask the male clerk or mechanic a question, I get the sense he doesn't believe I'm actually the farmer. He can be patronizing or, worse yet, flirts with me. I try to hold my own, but then I tend to come across too aggressive and, dare I say, 'bitchy.' Is it too much to ask to be taken as an equal and still be myself?"

Every time you strut your boots into traditional male farmer domain, you're cultivating change. Women may be some of the oldest growers on the planet, but you wouldn't always know that inside the walls of the FSA office or local feed store. Remember, we're still plowing new territory in modern times and, in many ways, launching ripples of social change every time we interact with a male farmer who is not used to dealing with a female on an equal basis.

Harvard social psychologist Dr. Amy Cuddy studies body language and, more specifically, how our body language can influence how we feel about ourselves. She gave a provocative TEDx Talk, "Your Body Language Shapes Who You Are," exploring how women tend to fall into patterns of powerless body poses, literally closing up and making ourselves small.

"Our nonverbals govern how we think about ourselves," Cuddy explains. While most of her research work focuses on how our body language influences how others perceive us in traditional situations like a job interview or class discussion in an MBA program, her insights and recommendations apply just as aptly to the male-dominated world of agriculture and conversations at the feed supply co-op.

What should we do before we approach our conventional farmer neighbor about his spray drifting onto our land or negotiate price on a used walk-behind tractor? According to Cuddy, hide out in a bathroom stall or another private spot like the barn and do two minutes of "power posing." Really.

"Power posing" entails taking on a position of dominance such as

- standing tall with hands on your hips (aka the Wonder Woman)
- sitting with your boots on the table, stretched out with your hands behind your head
- standing and leaning on a table
- sitting and leaning back in a chair, arm lying on the back of the chair next to you (i.e., taking up space).

Her research has shown that when women go through this simple two-minute ritual, our testosterone increases by 20 percent. Testosterone is the dominant hormone found to a greater degree in men but also found in women. Testosterone sometimes gets a bad rap, as we tend to categorize it as that uber masculine stereotypical hormone, something we women want nothing to do with. In reality, we have this hormone too, and these power poses enable us to control and manipulate testosterone to our advantage.

Cortisol, on the other hand, decreases by ten percent, which is a good thing. It's called the "stress hormone" and in larger amounts can inhibit our ability to function and lead, causing effects such as increased blood pressure and impaired performance. It's a good thing, then, that these power poses keep our cortisol lower and keep us calm under pressure.

I admit, this whole power posing thing seemed a little forced and silly when I first tried it. But over time, it felt more natural and I did feel more confident when walking into a stressful or intimidating situation. I get jittery before presenting to a new group of people, and I do my Wonder Woman in the bathroom right before starting. Indeed, it makes a difference. This has to do with what Cuddy refers to as "fake it till you make it," the idea that how we handle our bodies can change our minds regarding how we feel about ourselves.

Idea Seed:

"I've learned that our questions are more than valid. They are challenging and transformative. When you get weird looks or discouragement, you may be talking with someone who's never considered the alternative. That does not make you wrong. Don't give it another thought. Don't give away your power."

— Patty Cantrell, Principal Regional Food Solutions, Humansville, Missouri

Fitting in

"I'm used to being in the minority as a lesbian woman, but when I bought my farm and moved out to the country, that took feeling isolated to a whole new level. I realize in many cases I may very well be the first gay person someone around here has ever met. On one level, that can be a huge opportunity for me to educate and open people's minds and hearts. On the other hand, how can I find other women with whom I can kick back, who respect me for who I am?"

We bring diversity to the table, that's for sure. From the color of our skin and our cultural roots to our sexual preferences, we're breaking the mold of both traditional "female" and "farmer." In other situations, we may be returning to the roots where we grew up but coming back with new ideas like stewardship and sustainability that may not be the local norm. Where 2,4-D and Roundup flow like punch at a contra dance, how do we stay true to our vision, yet feel a sense of belonging in our new home?

I just need to wander up the road to Lindsey Morris Carpenter for advice. With her mom, Gail, she runs Grassroots Farm, a diversified operation of certified organic vegetables and pastured livestock. Morris Carpenter applies organics, sustainability, and community in everything she does. Resourceful,

Lindsey Morris Carpenter of Grassroots Farm.
ANNALISE CANFIELD

she's the one who can take barn scraps and MacGyver a chicken coop that even the Big Bad Wolf can't blow down. Such commitment and creativity rightly earned her the first John Kinsman Beginning Farmer Food Sovereignty Award from Family Farm Defenders. Not too shabby for someone who started out a decade ago in Philadelphia with a fine arts education, never having planted a seed.

"Queer, rustic, butch, playful womyn is how I'd describe myself," she says. From the short buzzed hair to 12 tattoos and counting, Morris Carpenter stands strong and proud of who she is and even more so of what she grows and raises on her farm. Yet she manages to successfully honor this genuineness and sense of self while living in a conservative and conventional farming community. Mine.

What would a queer gal do? She does 11 things, according to Morris Carpenter. Here are some of her tips on fitting in by not selling out:

1. **Don't assume anything about anyone.**
   "I am constantly surprised at who warms up to me, shuns me, or seeks me out. Give folks the same chance you would like to have," shares Morris Carpenter. "Remember the worst-case scenario is you don't talk, which is always their loss. I find it surprisingly easy to avoid folks who would rather avoid me."

2. **Be a good neighbor.**
   "A crucial key to happiness is being a good neighbor. Call around when you see livestock and pets outside of fences; maintain your fences; share your garlic and potato seed; and always invite neighbors to parties and events even though they may not show. Even if your lifestyle, gender, and farming operation are sheer opposites, we still have our physical location and appreciation of nature in common. And that's big."

3. **Sport your country creed.**
   "Learn the right vocabulary! Know the difference between hay and straw or a pig and a hog. Develop your own country wave and use it often when

driving by other cars and folks in the field. Smile in passing. Simple, but it goes a long way."

4. **Be prepared to talk shop.**

Morris Carpenter offers these reliable conversation starters: weather, equipment, wildlife sightings, and country hobbies. Stay away from the obvious — religion, politics, and social issues — until you are friends, or maybe never.

5. **Practice tolerance and patience.**

"Remind yourself that you may be the first person your neighbors met who is like you. Eventually they may feel comfortable asking you questions about your lifestyle, look, or unconventional choices. Or, like many of my neighbors, I accept the 'Don't ask, don't tell' policy. After all, I'm equally uninterested in their bedroom life and personal appearance choices. I hear homophobic and racist comments sadly too often, and pick and choose my battles. I am not trying to bring anyone into this century as much with my words as with my success and relevance as a queer organic vegetable and meat producer. Lead by example."

6. **Try to participate in local gatherings.**

Morris Carpenter admits she sometimes needs to force herself to go to a neighbor's event, realizing it won't be her tribe. But she always goes. "It's good to bring a buddy," she recommends. "Just because the host is cool, it doesn't mean the rest of the party are. Sharing a Busch Lite and bringing homemade barbeque sauce can open a world of opportunities when you least expect it."

7. **Engage in local trading.**

Swapping and trading are lost arts and a great way to connect to community. Morris Carpenter organized a "Tool & Supply Spring Swap & Trade" this past spring, during which local farmer neighbors stopped by and brought things they no longer needed to give away, trade, or sell — all shared over a potluck.

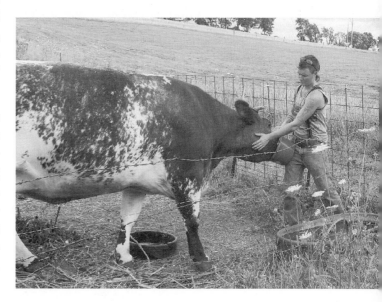

Lindsey Morris Carpenter and farm friend.
JENNIFER BOWEN

8. **Give country music a chance.**

"Seriously, you probably will have more pop country stations than any other genre available, so have an open mind. I can't count how many people I have a country music bond with and, well, nothing else."

9. **Be a good networker.**

Did you find a good small-square straw source? A good mechanic? Someone with a flatbed trailer for rent? It's a lot easier to fit in and be accepted if you are useful and looking out for your friends and neighbors. An added bonus is you keep that cash local.

10. **Be proud!**

"This is your chance to build your dream. Queers, women, and minorities in traditionally white male occupations are an integral part of the future of our social and food systems and survival. I'll say it again: Lead by example."

11. **Make your own community.**

"It's hard to feel alone in the country once you have found local people like you. Set up a reason for country queers and women in sustainable agriculture to meet casually."

Finding your local tribe

"How do you meet other women who share your passion for food and sustainability? I'd love to simply connect with others like me, but I'm not sure where they are or even if there are any of my kindred spirits out here. Traditional groups like women's clubs aren't right for me. I'm starting to feel lonely and afraid I'm going to turn into that crazy farmer lady with lots of cats all by herself in the sticks. What can I do?"

"I wouldn't be here today if it wasn't for the support of other local women farmers," shares Lori Stern, a local friend who moved to our area in 2010 with her wife, LeAnn Powers. Now successful local food entrepreneurs, Lori and LeAnn first launched Lucky Dog Farm Stay, which included converting their barn into a yoga studio where Lori teaches classes. The next entrepreneurial chapter involved starting Cow and Quince in the small rural town of New Glarus, creating the Community Supported Restaurant (CSR) and local food hub in our county. "When I needed anything, from contacts for suppliers to advice on dealing with state inspectors, I knew who to talk to. That kind of

support and knowing someone is on your side is priceless, especially in our traditional rural community where male-run, conventional agriculture is still the norm."

"Build it and they will come" may work for a baseball field of dreams in the middle of Iowa, but it also adds up to good advice when you harbor a need to connect with other local women. Start initiating, keep inviting, and stay in it for the long term and you will amaze yourself with the women who show up around the table, thanks to your leadership.

The roots of my local tribe go back five years to a November evening in Madison, Wisconsin, when I taught an introductory beginning farming workshop for women via the MOSES Rural Women's Project. About 30 women gathered that night, none of whom I knew. After a brief welcome, I kicked things off as I always start my workshops, with everyone introducing herself: name, where she's from, and what farm dream brought her here that night.

First, Lindsey Morris Carpenter stood up and introduced herself as running Grassroots Farm near Monroe. Later, Lori Stern gave a shout out of how she and her partner had just moved to the area with visions of launching a farm stay. The introductions wrapped up with Katie Lipes, a chicken-raising momma with a baby

Idea Seed:

"Race continues to play a big part in our food system, so these barriers need to be broken so we have a more holistic approach to what we're doing. Change is scary. Our population is becoming more brown and there needs to be ways where we can together sit at the table and talk about the way decisions are made. Food should be a right for all and not a privilege for some. When a person is given dignity when receiving food, it will thwart the shame associated with hunger and poverty."

— Karen Washington, founder of Black Urban Growers and farmer and co-owner of Rise & Root Farm, West Lebanon, New York

Karen Washington, farmer and co-owner of Rise & Root Farm.

ETHAN HARRISON

Idea Seed:

"The support and sharing of information that women farmers are willing give to each other is inspiring. We'll meet for a gathering at a local farm and I am always amazed to see how camaraderie and connection through the hard work of farm life lifts them up. The network gives these women the knowledge that they are not alone, even if they sometimes feel isolated on the farm."

— Melissa Fery, OSU Extension Service and coordinator for the Willamette Women's Farm Network, Corvallis, Oregon

in a sling. I thought, "Who are these cool women, located within 30 miles from me, and why haven't I met them before?" That intrigue lingered as I drove the hour home, up and down the hilly rural roads. Somewhere amid those cornfields, I decided to throw a potluck.

I emailed an invite for the first Sunday night in December and left it at that. And come together we did. As the homemade hooch flowed and the artisanal goat cheese and cracker tray got passed around, a welcoming warmth filled the room. Early in the evening, folks already started shouting out, "When are we doing this again?"

Today, our South Central Wisconsin Women in Sustainable Agriculture group boasts over 100 area women who gather at six on-farm potlucks throughout the year. What continually amazes me is the growing list of tangible outcomes of women informally but regularly gathering over supper. A beginning farmer connected with a woman with extra land to lease, and a partnership formed. Some women started a chicken-feed-buying co-op to enhance buying power. Countless baby goats, heritage hogs, and local insurance agent recommendations are shared.

These outcomes go beyond the sharing economy and spark new businesses and dollars flowing into our community and local economy. For example, Anna Landmark and Anna Thomas Bates met at a potluck and eventually formed a strong business partnership. Landmark was already on her way to earning her cheesemaking license to launch her own operation but needed a partner to help with the business and marketing. Bates, a savvy food writer, gladly filled that role. The duo launched what is now an award-winning cheesemaking venture: Landmark Creamery.

"We're both moms with kids in the same school district, but we never met until these women-in-agriculture potlucks," reminisces Landmark. "Even if we had met in a school setting, I'm not sure we would have had the opportunity to connect in a way that we did over cheese and wine. The potluck provides a welcoming, supportive setting through which women like myself and Anna feel comfortable sharing our big-picture visions and dreams."

Another group of women who connected at these potlucks started a local chapter of the Wisconsin Farmers Union with a more rural policy and grassroots organizing focus. Our big public tourism event each August involves over 20 women-owned farms giving tours, workshops, and on-farm culinary events. It's called Soil Sisters: A Celebration of Wisconsin Farms and Rural Life (www.soilsisterswi. org). What started as a potluck turned into a farming and food celebration to boost the local economy by tens of thousands of dollars.

Here are four tips to get you started with a potluck-based local network, drawn from our experience in Wisconsin:

Soil Sisters of Wisconsin.
JOHN D. IVANKO

Soil Sisters logo designed by Brett Olson, Renewing the Countryside.

1. **Commit to lead for one year**

   Make organizing a local network your personal project for at least one year. While these networks do not need to be formal organizations with elected officers, bank accounts, and high commitments, they still need consistent leadership to get successfully off the ground. Be that person to recruit other potluck hosts, set a schedule, send out reminders, and answer questions.

2. **Set calendars and consistency**

   Given the demands of the growing season, it helps to set potluck dates for the whole year in January. This way, it's much easier for hosts to commit before their schedules fill up.

   Our gatherings are usually a Sunday supper but sometimes brunch or lunch. Our format is typically the same. We start with a farm tour followed by an activity, and then we do introductions and eat. The activity could be anything from Anna Landmark giving a mozzarella-making demo to Lori Stern teaching a yoga class. At one potluck we did a "fermentation show and tell" because our group had both seasoned fermenters and beginners wanting to learn. The experts brought jars of their latest ferments and shared how they did it; questions and sampling continued until we all had our fill.

We always have one co-ed summer gathering where women bring their spouses, partners, and kids. The annual event balances the women-only aspect of our other gatherings as it gives everyone a chance to meet the men and kids in our lives.

3. Create communication channels

Create an easy way for folks to keep in touch, based on the media preferences of your group. Some like a Facebook page, but for us, a simple free email forum works well and serves as a means to find a home for extra kittens or to share information on a community event.

Prairie walk at South Central potluck with April Prusia of Dorothy's Grange. LISA KIVIRIST

4. Keep it local

We have a Yahoo email forum that isn't open to the public. Requests come through me. Our goal is to develop a tight local network of women who know each other individually. Therefore, you need geographic ties and a kinship with shared values of sustainability and local food. Some women on the list may be looking to move to the area, hoping to make connections and find local resources before they buy a farm.

Integrating family and kids

"I want to have kids someday but am worried about how to integrate and balance that with the business. How can I be a mom while farming?"

At a MOSES farming conference recently, on a whim I organized a "farmer roundtable discussion" on this topic of integrating family into the farm business. This wasn't a workshop, but rather an informal discussion. I was expecting maybe ten people to show up and ended up with a crowd of 40, mostly women and a few men, sharing a common interest in finding that holy grail of on-farm parenting. What's that magic formula to make it click?

"Raising kids on a farm creates the most joyous, crazy, unpredictable, rewarding life ever possible," shares Dela Ends of Scotch Hill Farm in Brodhead, Wisconsin. I roped Ends, an organic farmer for over 20 years, into leading this small-group discussion with me because she and her husband, Tony, raised four kids while they grew their CSA. "Don't worry about trying to figure it out or feel guilty about messing something up in raising your kids. A farm is an amazing learning laboratory for children and you're providing a strong role model for them. Every day they see their mom out there in the field, doing what she loves."

Farming gives you a wide forum to create a situation that best suits you and your family, one that prioritizes your role as mom. This freedom stems from being self-employed. You can call your own shots and have greater flexibility to maneuver your business to suit your family needs.

In theory we might call the shots on our farm, but integrating kids doesn't always neatly blend in. Storms roll in and goats get out of the pen. Your youngest child catches the flu and you're up through the night before market day. The unexpected happens and we need to continually adjust and tweak our plans and expectations accordingly. Life happens.

Following are some advice tidbits from seasoned mommas in four key family integration categories. Their advice stems from discussions at conferences and at various women farmer workshops I've facilitated over the years:

Get creative with little ones

You love both your farm and being a mom but blending those two roles can feel like the ultimate challenge on many a day.

"Realize that sometimes you need help to make it work; an extra hand goes a long way," shares Kat Becker of Stoney Acres Farm in Athens, Wisconsin, the pizza-farm running, CSA-managing mom we met in Chapter 6. "Barter for the services and extra hands you need." Becker trades a CSA share with a woman in the area who comes and watches her three young kids once a week. "When I don't have to keep an eye on my kids, I focus and efficiently crank on, completing farm work in the time I have." Becker barters a CSA share with another local woman who helps with housecleaning during the peak summer season.

If other nearby family farms are also juggling young kids, perhaps you could work out an informal "childcare co-op" where you take turns watching the kids for one morning a week. Some women can manage holding their baby in a sling or backpack while working.

As kids grow old enough to play independently under your watchful eye, the new challenge may be to balance the attention you give them with the

competing work demands of the farm. "Finding a special farm ritual that you do daily with your kids gives you some anticipated time together that shares the work you're doing in the field," offers Dawn Combs of Mockingbird Meadow Honey and Herb Farm. "My kids are both under five and we regularly have a tea-making time together. We go into the gardens and harvest the herbs they want to use and brew up some 'fairy tea' as we like to call it."

Sometimes a small, daily treat can go a long way in occupying young children. "When our kids were little, we'd pack them a lunchbox of snacks and beverages in the morning and this would be their personal food stash throughout the day," shares Atina Diffley, an award-winning, long-time organic farmer in Minnesota. "They looked forward to their filled lunchbox. This also took care of the kids getting hungry and thirsty and me needing to break from work to get them something." It also increased Diffley's kids' independence and engagement with the farm. "By the time they were four they often packed their own lunchbox to 'go to work.'"

Explore homeschooling

I first encountered homeschooling at organic farming conferences years before our son was born. Something about how these kids carried themselves, the poise with which they could speak to adults, and their curiosity for the world around them impressed me. When I had our son, Liam, in 2001, I started going to LaLeche meetings and immediately connected with a group of like-minded families in our area.

Many of these families chose to homeschool as a means of exposing their kids to a world beyond the classroom and curriculum requirements, a philosophy I found appealing. We decided to give homeschooling a try when Liam would have started first grade, and we quickly found this approach blended well with our farm lifestyle, including traveling off-farm during the winter months.

We're not alone in our homeschooling journey. According to the U.S. Department of Education, approximately 1,770,000 students are homeschooled in the US, a 17 percent increase between 2002 and 2007 that adds up to 3.4 percent of the school-age population.

"There's no shortage of authentic learning opportunities for homeschooling on the farm," shares Laura Endres, the fifth-generation involved with her family farm, Irish Grove Farms, raising grass-fed beef in northern Illinois. "From animal care to respect for the environment, from manual labor to cooking and handiwork, your farm offers an engaging platform to learn the essentials of a good life." Endres is also a national unschooling advocate and speaker, a personal trainer, and a co-founder of the Walk Fit Challenge, an online fitness company.

"The biggest challenge of blending homeschooling and farm life is there's always work to do and it's easy to feel you are never done," adds Endres. She and her husband, Rob, homeschooled their two sons. Her solution? "Resist the need to see tasks as 'done,' but rather all elements as part of the process.

Chores never go away; you're simply somewhere on a continuum." Make room for downtime and field trips and playgroups.

Technology and high-speed Internet expand rural homeschooling options tremendously. Frankly, I'm not sure how I'd homeschool without these resources as my son, Liam, is quite the computer geek. Even in our rural homeschool setting with very limited expertise ourselves, we can readily support his passion for technology. Sometimes Liam teaches himself, like learning to code through free online resources. Other times, John and I support him by connecting him with opportunities we learn about online. Liam applied to serve as a "kid reporter" for *Time for Kids*, the classroom version of *Time* magazine. Through this media credential, he and John attended the Consumer Electronics Show (CES), the annual mega-event of all things tech. That gathering alone accelerated his learning curve in ways a classroom never could, as well as further connecting him to new ventures. In the lunch line at CES, Liam met an editor for *Innovation & Tech Today* magazine and now has a regular, paid writing gig.

Technology and the Internet can also open up social channels for homeschooled rural kids. I often am asked if Liam feels lonely or lacks social skills because he's an only child homeschooled on a farm. Once the questioners meet Liam, they quickly realize that his social skills shine, perhaps because he's had ample practice with a steady stream of B&B guests over the years. But social skills with his peers come via Skype, as just about every night a dozen or so friends from across the country dial in and talk.

For a while, I admit, I was Skype skeptical. It felt foreign to me, who prefers face-to-face settings. However, I've warmed up over time and now champion it as a means for teens to connect. He's in a safe environment on private servers where we know the other kids involved. The dialog is no different from when I was a teen, except that I was relegated to talking to one person at a time, dragging the corded phone into the closet. Liam can connect with friends

Idea Seed:

"I became a farmer after a mid-life career change, shortly after my middle son was born. I think raising my kids on the farm has been a great environment. The most important lessons learned from seeing their dad and I run our farm and business every day are 'There is always a solution' and the corollary, 'If you can't find the solution, find someone who can help you.' Those two principals have helped us succeed in the always-challenging and ever-changing world of the farmer entrepreneur, and I think they will serve our boys well in whatever path they follow."

— Beth Osmund, Cedar Valley Sustainable Farm, Ottawa, Illinois

easily and often in this setting, something that would have been impossible just a few short tech years ago.

Foster young farm entrepreneurs

The farm provides a safe and accessible venue for kids to build their entrepreneurial skills and muscle. They learn from you in your role as business owner and can, if you support it, develop the ability to create a mini venture of their own from the ground up. We've never given Liam an allowance but always helped him get something going on his own. Every season he can try out a new idea. He's grown popcorn and basil to sell to our B&B guests, developed a knack for making balloon dogs, and worked farm events and other gigs for tips.

"My daughter, Selket, shines at the farmers' market where she sells her bread and toffee," shares Jane Jewett, owner of Willow Sedge Farm, a livestock operation in Minnesota. At 15, Selket has four years of experience behind her making her products and selling them at the market. "She's pretty shy normally, but through this experience she has learned that she can talk to customers and share her story. I also helped her set up her own checking account, as she needed a parent as a co-signer. This enables her to be responsible for her own finances and do her own transactions when she needs to buy flour and other supplies."

Beth Osmund of Cedar Valley Sustainable Farm.
JOHN D. IVANKO

From managing your tractor mechanic to chatting with your neighbor next door, from bonding with babies to finding female friends, it all boils down to relationships. As women growing food, we know health doesn't start and end with what's on the dinner plate. Health encompasses so much more, with a core element being the people we have in our lives. The next chapter takes being well a step further by examining how to craft a balanced life.

Chapter 10

Crafting Balance

For me, May 1 rings in the annual vernal flurry of spring, when everything feels infused with hope and renewed energy. We prep growing beds, plant seeds, clean out the barn, and run on high-octane energy as we anticipate the season ahead. Summer rolls in as I discover the first ruby ripe strawberry in late May. The quack grass and thistle have begun their advance into the fields, keeping us busy. The temperature rises to the pleasant upper 70-degree range and, for the moment, farm life feels both satisfying and manageable.

But by late August, my energy tank starts to run on empty. The humidity grows more unbearable along with that Eight Ball zucchini in the field that I keep reminding myself to harvest. By then, the lush green early summer landscape dries out to tones of yellow and brown, other than where we've irrigated. I stumble through September, and by the time the first frost hits in October, I celebrate that the growing game is nearing its end. I crave that night when it will be cold enough to linger around the woodstove.

Throughout winter I write, rest, cook, reflect, and catch up on house projects. When the bitter cold rolls in, my family and I escape to warmer climates to write, work, and keep our hands in the soil. By then, we're craving the warmth of the sun, green turf, and greens on our plate. By March, we await the first warmer day to get out and prune the raspberry bushes.

A farm-based livelihood celebrates the yin and yang of balance. Dating back to ancient China, yin and yang represent the belief that everything in the universe consists of two forces that are opposing but complementary. Boys

and girls. Cold and hot. Potato chips and chocolate. Who knew that the savory-sweet thing would be so good?

On the farm, yin and yang symbolize the slower dormant season balancing the chaos of peak harvest. Exactly how these highs and lows even out depends on the growing climate of where you live and what you can grow when. But everything has the potential to balance. We need to prioritize finding that sweet spot.

The quest for the life-balance sweet spot is what this chapter is about. This ongoing journey toward feeling in sync with the seasons, our farm and, importantly, our selves involves thought, time, and heart. After years working with women farmers and conducting interviews for various articles, I have yet to meet a woman who claims she's found that magic sweet spot formula and has it together. We're most definitely on an ongoing journey where balance is a goal, but we know it will never be the destination. We'll have moments like that June day when strawberry shortcake never tasted better alongside that summer night when a storm rolled in and the wind turbine blades cracked.

The journey toward balance remains a topic we relish discussing. We share ideas of what worked for us — and didn't. In this chapter, we'll look at self-care through managing our time and nurturing our bodies, along with multiple ideas from other female farmers on what works for them.

Time management

Time management. For those of you who, like me, escaped cubicle jobs for the dream life in the country, the words "time management" sound a little too reminiscent of careers and day planners. Didn't we triumphantly trade working for someone else for independence on the farm? Can't we call our own shots, do what we want to do when we want to, and bask in the sunshine of self-employment?

Yes and no. Most resoundingly, farmers should relish our autonomy. Yet, like a well-trained muscle, time management is an ongoing skill that needs to be developed. Why is this important? The to-do list never ends. With fieldwork, animal chores, and outbuilding maintenance, added to a lengthy list of other family and life responsibilities, farm life can easily spiral out of control and feel more like a burden than a blessing. Without time management we grow stressed and overwhelmed and lose touch with the everyday joys of farm life. Understanding yourself, your needs, and your work style enables you to best manage the job, work toward long-term goals, and enjoy the ride.

The ironic side of time management is that we often get too wrapped up in daily responsibilities to have any time to think about improving this skill. Here are five easy tips to amp it up:

1. Know thyself

The first step in effective time management is understanding yourself: How do you work best and in what situations? Take advantage of the autonomy that farming allows, and plan your daily schedule around when and how you work best.

Natural biorhythms play a leading role in understanding yourself. Are you a morning person or night owl? When does your energy level hit a high peak, and at what time of day do you crave a caffeine fix? Plan your day as best you can around these natural tendencies and work more efficiently.

I'm a definite morning lark and by 5 AM — sometimes even 4:30 AM — you'll find me at the computer. Why am I pecking on a keyboard versus plucking weeds during my peak time? I focus my best creative time on writing, taking advantage of the quiet early morning hours before my family wakes up. I don't need to be in full idea flow when I'm weeding or watering; I do those chores later in the day.

Another important aspect of knowing yourself is identifying your core priorities. This is what led us to decide to not have animals. For years we raised a small flock of free-range chickens for fresh eggs. During the winter, chicken responsibility conflicted with our travel. It placed too much of a burden on friends to stop by to check on the chickens; plus tensions started escalating with our neighbor because of their roving dog picking off our flock. Add to this situation that I could easily purchase eggs and support a local friend's fledging farm business. So we gave away our flock and eliminated that time commitment, not to mention stress.

2. Partner with the seasons

Mother Nature gifts us with that ultimate time management tool: four seasons. Align with the natural ebb and flow of the seasonal calendar and take advantage of inherent busy peaks and restful valleys. Summer may bring a chaotic climax of craziness but that balances with the slower, reflective winter season. Focus on the essentials that need to get done in the summer and "back burner" other projects till winter.

For example, I'm always aiming to simplify summer produce processing in our home kitchen. I'll quickly tray freeze clean, whole tomatoes, placing them in freezer bags after they are frozen hard. During the winter, when it's nice to have something slowly simmering on the woodstove's cooktop, I defrost the tomatoes in the refrigerator, peel them (super easy with frozen tomatoes), and add seasonings. The tomatoes, garlic, onions, and herbs slowly fragrantly thicken into sauce, which we use for pizza and spaghetti.

A seasonal time management approach proves to be the antidote for perpetual procrastinators. Before my farm-based livelihood, I tended to be queen of last-minute deadlines. Now that I understand the farming workload, I know I won't have time to write an article in June or create that photo collage for my dad's July birthday, so I'm motivated to wisely use my winter time and work ahead. Winter gives me time to do my annual clean out and reorganization of the home office and attic.

Additionally I use daily weather forecasts to plan weekly schedules. Rain predicted tomorrow? Hang the laundry out today. Rather than weed during uncomfortably hot and humid days, work in the kitchen or other indoor projects during those times. Get outside when the mercury drops to a more comfortable zone. Weave in a work schedule that's easier on your body.

3. Prioritize

Lisa painting the Inn Serendipity B&B sign.
John D. Ivanko

Accept that there's only so much you can get done in a given day, and prioritize. Need a motivational boost on a project that simply must get done? Break it into small chunks.

For example, the paint on our wooden farm sign kept peeling off and looked awful, but I kept getting caught up in other farm projects and couldn't block out a sunny afternoon to paint the whole sign. With a closer look, I realized only the orange and white paint were peeling; the blue sections looked fine. I located and organized my paint, brushes, and sandpaper. The next day I quickly sanded and painted only the orange and white sections. The whole process took less than an hour when I broke it down.

4. Pick low-hanging fruit

Feel free to take the easy way out, looking for the low-hanging fruit that provides the same juicy flavor without needing to wobble on a high ladder. Don't feel obligated to take the traditional long road if there's an easier way that produces similar results in less time. This might mean breaking with your own expectations and acknowledging that you don't have to do everything yourself. Hiring someone, especially for that odd job on your to-do list that isn't getting done, is perfectly acceptable.

"Always weed long rows backwards so you can see what you already accomplished rather than looking at what's left to do and get discouraged," advises Dela Ends. "This applies to the growing fields as well as life. Celebrate how far you've come."

5. Use shortcuts and systems

Farmers and lists go together like toast and jam. Create both master checklists ("everything I need for the market booth") and reminder notes after each season for the next round ("bring extra tablecloths; explore using chalkboards for signage"). My lists multiply in topic and grow in length so much that I carry a clipboard. A little nerdy, no? My family chides me about looking like a camp counselor.

"Getting organized during the winter months is a huge part of me operating efficiently during the summer," shares Debra Sloane. "During the winter, I order my tools and seeds and create a detailed schedule on a spreadsheet of what I will be doing every week. When the craziness of summer hits, I go on autopilot and follow my schedule and jot down some notes for next year as I go. As a woman farming solo, this routine helps me prioritize and focus."

Idea Seed:

"Be efficient with your time when it comes to solving a farm challenge. If you have a problem, chances are another farmer already solved it. I find farmers in Eastern European countries or Mexico and Central America particularly inventive as they have less cash and fewer purchase options and have to create many items themselves. Type in your issue you are having into Google Translate and translate it into a language like Spanish or Ukrainian or Slovakian. Plug that word into YouTube and you'll see videos pop up of what farmers in those countries have done. You may not understand the language, but I find the visuals alone very helpful."

— Clare Hintz, Elsewhere Farm, Herbster, Wisconsin

Tool Shed:

Farmer Chick Lit Reading List

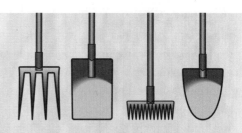

For some female farmer R & R take a read of these fabulous farm memoirs exclusively written by female farmers, capturing the empowering stories behind personal journeys on the land. Spoiler alert, all have happy endings!

Turn Here Sweet Corn: Organic Farming Works, by Atina Diffley

The Seasons on Henry's Farm, by Terra Brockman

Hit by a Farm: How I Learned to Stop Worrying and Love the Barn and *Sheepish: Two Women, Fifty Sheep, and Enough Wool to Save the Planet*, by Catherine Friend

One-Woman Farm: My Life Shared with Sheep, Pigs, Chickens, Goats, and a Fine Fiddle, by Jenna Woginrich

Farm City: The Education of an Urban Farmer, by Novella Carpenter

Shepherdess: Notes from the Field, by Joan Jarvis Ellison

The Dirty Life: A Memoir of Farming, Food, and Love, by Kristin Kimball

The Feast Nearby: How I Lost My Job, Buried a Marriage, and Found My Way by Keeping Chickens, Foraging, Preserving, Bartering, and Eating Locally (all on $40 a week), by Robin Mather

Fifty Acres and a Poodle: A Story of Love, Livestock, and Finding Myself on a Farm, by Jeanne Marie Laskas

This Organic Life: Confessions of a Suburban Homesteader, by Joan Dye Gussow

A Country Year: Living the Questions, by Sue Hubbell

Yin and yang of controlling chaos

After a full day of chomping away at that to-do list, you need a little R&R time. Likewise, at the end of the growing season after you pull in the last leek, take advantage of those upcoming dormant months to slow down. Caring for ourselves, both physically and mentally, forms the best insurance for us to keep farming long term.

We sisters know how to work a healthy, balanced farm life:

"I find it helpful to learn from other professions when it comes to self-care," shares Katie College. "One challenge is that, living in your workplace, it's so easy and sometimes unavoidable to work from dawn to dusk. That can drain you mentally and physically. If we worked in say, advertising, and put in the same hours with the same intensity as we do in farming, we'd realize that we're putting our bodies, minds, souls, and relationships at risk.

"It sounds silly, but I dress for work as if I were getting in the car and going somewhere. I've stopped running out in my jammies to feed the chickens. When I wear my work 'uniform,' I'm working. When I'm done for the day, even if it's very late at night, I'll shower and change and have a glass of wine to signify the end of the workday. This has helped my relationship with my husband as well, as he does not work on the farm. This defines time for 'us' that's separate from the farm."

Relinda Walker plans a major trip off farm for about a month every year,

Relinda Walker
John D. Ivanko

at the slowest time of year. "For us here in southern Georgia, that time comes in August, when the weather here is so hot and humid that we don't have much in the fields. . . . Seeing other places and surrounding myself with a totally different growing climate rejuvenates me, and I make a point to let go and disconnect from the day-to-day farm operations. For most of the year I run by my to-do list and carry my clipboard everywhere, but I make a point to not do that when I'm away. It's healthy and grounding for me to step away and disconnect. I come back recharged and recommitted to my farm."

It's a 45-mile drive to the nearest city for Steph Larsen, organizer for the Sierra Club's Beyond Coal Campaign, and beginning farmer in Saint Ignatius, Montana. "Because I value sustainability and pay attention to my carbon footprint," she says, "I used to feel guilty every time I drove.

Then I realized that if I don't take care of myself — for me that means visiting friends, seeing a film, going out to dinner, being in a community — then I get depressed and I can't do the good I was meant to do in the world. So I set aside the guilt and try to plan trips well so I can accomplish multiple things. I am healthier for it."

For Dela Ends, the kitchen is a favorite spot for inspiration and her hide-out from outdoor farm work. "It may sound counterintuitive, but even after a day of harvesting, cooking that harvest is very relaxing for me," she shares. "Canning and putting up produce, both for my own family and to sell under our state's cottage food law, is something that I seem to have energy for, maybe because seeing those jars on the counter is so rewarding. A glass of red wine at the end of the day signifies a shift from the field to indoors and celebrates a day well spent."

For Maisie Ganz and Willow Hein, of Soil Sisters Farm in Nevada City, California, relaxation is a trip down the road to a swimming hole on the Yuba River, affectionately called the "farmers' hole." Two or three times a week they hang out there in the late afternoon. "The water rejuvenates us and we relax on the round granite boulders. Take time for yourself and prioritize these rituals, especially if you are near water."

Food as fuel

"Prioritize so that you get your personal share of the best of what you are growing first," advises Dawn Combs. "Savor and enjoy your harvest." Combs learned this first-hand when she realized she rarely savored her own honey. "Everything went toward sales, and I wasn't taking the time to connect and enjoy what I created."

That can be a challenge when the summer chore list grows long, and time in the kitchen gets clipped by needing to clean the stalls. Have a plan and be prepared. I write out a menu list for the week, which I hang on our refrigerator. This ensures I cook around what's abundant and fresh while encouraging me to peck away at the food preparation. If I have some time in the evening right before I go to bed and notice we're going to have zucchini feta pancakes for a B&B breakfast the next morning, I'll shred the zucchini then for quicker prep the next morning.

Our farms express the manifestation of our lives, our passions and vision. Maisie Ganz and Willow Hein took that "manifest" idea one poetic step further by creating a personal "Working Manifesto" for the farm. Author Ganz shares her words that she and Hein work, play, and live by, inspiration we can take to heart in our journey toward balance:

A Working Manifesto

To create an entity that can give us the strength, validation, and support we need.

   To pursue farming by our own standards of success and health. To value women's voices and women's wisdom. To celebrate the slowness of seasons, the patience of god, the beauty under our feet, between our hands, and in the river. To lift up song, dance, prayer, reflection, celebration, communication, growth. To let ourselves be overwhelmed with delight and joy, and overwhelmed with heartache and loss, and to hold space for both. To hold each other up, to hold each other accountable, and to honestly and gently express our needs. To daily affirm one another, and to appreciate the presence, heart, and skill each woman brings. To recognize our different paths and goals, and to revel in our ability to weave them together. To lead with confidence and flair. To do what feels right. To change our internalized doubts and pains into healing and positive action. To inspire and empower other women in various facets of their lives. To include men to participate and share in our vision. To take the time to enjoy the fruits of our labors — cooking slow dinners, eating carrots in the field, and drawing, painting, and photographing what inspires us. To dig, cultivate, seed, transplant, sweat and move with intention and self-awareness. To value and love ourselves deeply. To practice self-care. To practice kindness and compassion. To commit to living fully and joyfully.

Maisie Ganz and Willow Hein of Soil Sisters Farm, Nevada City, California.
Stephanie Brown of Peaceful Valley Farm Supply, Grass Valley, California

Kriss Marion, Circle M Market Farm, Blanchardville, Wisconsin

Farm Girl Boot Camp:
A Forty-something Farmer Finds Fitness

"Something clicked with me when my last child fledged the home and left for college," recalls Kriss Marion, who runs Circle M Market Farm, a diversified small-scale operation in southern Wisconsin with vegetables, meat, and, now that some space has opened up in the house, a farm-stay bed and breakfast. "I had been running the farm for the past eight years since we moved to Wisconsin from Chicago. My husband, Shannon, travels a lot for work, so my life has been about the kids and the farm. All of a sudden, my time opened up in ways that I could prioritize and focus on my health, both physically and spiritually."

Kriss Marion of Circle M Market Farm.
WWW.SARAANNAHANSEN.COM

A first area of focus for her was food and fitness. Marion was an avid athlete in her younger years. Although she kept active with farm chores, as with many of us, the pounds slowly crept on and overall fitness went down. "Not only that, but even though I was growing organic vegetables for other people, my days were so full and so long I often threw together lunchmeat sandwiches for the crew at lunch and ate nachos from the snack stand at one of my kids' sporting events for dinner," admits Marion. "When my last child went to college, I took a deep breath and realized it was time to take care of me."

Some might suggest moderation and gradual change, but that's not Marion's style. "When I commit to something, I'm fully in," she says with a laugh. On April 1, 2014, she rekindled her love of fitness and began tracking

her calories and walking daily in the woods next to the farm. By April 1, 2015, she had lost 50 pounds and now teaches a 4:30 AM boot-camp-style fitness class at the high school gym in town. Rebooted and re-energized, Marion offers three tips on staying fit:

1. **Track your food**

   "There's been tons of research studies that show writing down what you eat is a major success strategy in losing weight and eating consciously, but it wasn't until I tried it myself that I realized this," Marion shares. "I had no idea the amount of calories in some of my favorite foods; nor did I realize the quantity I ate." Today there are accessible apps and free online programs to easily track what you're eating, both calories and nutritional counts. "I use WebMD, which takes me about thirty minutes a day to enter what I ate, how much I exercised and how much water I drank. It's a commitment, but I find it keeps me conscious of what I'm eating and, importantly, the amounts. There's no lack of good food around here, that's for sure."

2. **Prioritize a daily workout**

   "I used to think that having a farm was a substitute for working out, but it simply is not," explains Marion. She also thought it was normal to have back, knee, and foot issues all season long. "The older we get, the more important core work and stretching become, as this keeps our joints nimble," explains Marion. "It doesn't matter what you do as long as you work out consistently. Pick a time during the day — perhaps first thing in the morning or at the end of your day — and take care of yourself for fifteen minutes minimum. Move your body in ways that feels good to you."

3. **Find new passions**

   "Now that I didn't need to drive my kids around or attend lots of school sporting events, time blocks opened up for me I hadn't had for years," adds Marion. "Open time enabled me to get more active locally with my fellow farming and local food community." The National Farmers Union proved to be the launch pad for Marion as she attended various leadership trainings, particularly focused on grassroots organizing in rural areas.

She started a local chapter of the Wisconsin Farmers Union, which brought together people, mostly farmers, committed to the same ideals. This gives her both a renewed local tribe of allies and a direct connection and motivation to creating change in her community.

Supporting other female farmers in body care is another of Marion's passions. She set up an informal "farm girl boot camp" Facebook page (facebook.com/farm girlbootcamp) where she shares fitness and health tips and provides a forum to set goals in a supportive — but accountable — environment. Her motto: You first, then the farm.

Epilogue:
Plate to Politics — Lead the Change You Seek

SEED PACKETS SHOULD COME WITH WARNING LABELS: The more we plant, the more we nurture those tiny seeds into zucchinis the size of baseball bats, and the bigger we dream. We see beyond the property lines of our own acreage and feel connected to our larger food system, realizing the many and mighty challenges, discrepancies, and barriers that currently exist.

Why can't all children, particularly those in urban communities of color, have easy and affordable access to the kind of healthy food I raise? Why does my neighbor growing conventional corn receive over $80,000 in federal subsidies? Why do I never see my elected officials championing policies that prioritize land stewardship? Why does the USDA still classify fruits and vegetables as "specialty crops?"

It's easy to stop at simply asking the questions. It feels therapeutic to vent and lament with fellow female farmers over a glass of wine about all that's wrong with our food system and what we would change if given that magic wand. But then we wake up the next day to the status quo. We continue to live on our "happy island" of organic agriculture, surrounded by a community of like-minded spirits who share our worldview. The challenge remains to cross the border, connect with issues and people outside of our sustainability sphere, and use our farming stories and authentic voice to collaboratively move the dial to lasting change.

Lasting change. That's the message of this final chapter. So far in this book, we have focused on developing our farm business and skill strategies. Let's now move to growing things beyond our field, move beyond our individual plates to politics, leadership, and strategies in order to both leave our communities with cucumbers and create forward momentum to be that change we seek.

State of the field

Although progress has been made on the women's leadership front, we still have a long way to go toward equitable representation. According to the Center of American Women and Politics (CAWP) at Rutgers University, women hold 19.4 percent of the 535 seats in the current 114th US Congress. While this represents a slow but steady increase since 1917, when Jeannette Rankin of Montana was elected as the first woman to serve in Congress, our voice is still not equitably represented at the decision-making table, considering we make up a tad over 50 percent of the population.

Women of color add up to a mere 6.2 percent of Congress but constitute approximately 18 percent of the US population overall and one third of the female workforce, per the Center for American Progress. The Center notes that this stands in stark contrast to many other countries: Finland, Iceland, and Norway lead the way, with 43 percent, 40 percent, and 40 percent of female legislators, respectively, in 2012.

"Research shows that women's leadership in various contexts adds transparency to the process, increases diversity and collaboration at the decision-making table and prioritizes the health of future generations," shares Liz Johnson, director of operations and community at VoteRunLead, a national, nonpartisan organization that unleashes the power of women leaders in democracy through training, technology, and community. With a strong passion for supporting rural women leaders and female representation in the agriculture world, Johnson helped launch Plate to Politics, a national collaboration that I and a group of female food system leaders founded to start providing leadership training and support targeting women working toward food-system change.

"It's hard to believe, but the research also shows that little girls still think that politics is for boys; we can't afford to continue to accept the rhetoric that politics is icky and something for other people," adds Johnson. "If we really want to change policies, both in this country and around the world, we can't afford to sit on the sidelines of our own democracy. Aligning our advocacy work with community leadership and the power to legislate on our values is critical to moving forward on policy. If not you, then who? Invite a woman to run for office, and remember that girls are watching for our lead."

Leadership ideas

However, just like our approaches to farming, women's leadership can take many forms. We don't just plant tomato seeds; some plant Black Krim, some Chocolate Stripes, and others Blondkopfchen. Running for office, particularly on a federal level can be a huge source of influence, but realistically that is not within the commitment or, frankly, interest sphere of us women farmers. So, with our hearts and hands in the soil, in what other ways can we integrate a leadership component into our farm plan mix?

Idea Seed:

"We live in such an extraordinary time, facing so many challenges. Some days it is difficult to see we are the solution. We are compassionate nurturers, sisters to Mother Earth, bold change agents, lovingly turning the soil of life, harmonizing with nature, not conquering it. We are capable of extraordinary acts of kindness. What better ways to change the world?"

— Theresa Marquez, mission executive,
Organic Valley Family of Farms, LaFarge, Wisconsin

The key here stems from the underlying theme throughout the pages of *Soil Sisters:* women cultivating new approaches to farm-based livelihoods. We increasingly see inspiring women in our food and agriculture community who play by their own leadership rules, latching on to issues and opportunities they feel passionate about, and creating avenues for change on their terms. Consider the following examples.

Atina Diffley: organic pioneering activist

Running one of the first certified organic farms in Midwest, Atina Diffley knows how to grow good food, but her impact yields far beyond the sweet corn harvest. With the threat of eminent domain for a crude oil pipeline proposed by notorious polluters Koch Industries, Diffley successfully led a legal and citizen campaign to halt the project. She shares her story through authoring award-winning memoir, *Turn Here Sweet Corn: Organic Farming Works.*

Marji Guyler-Alaniz: photographer on a mission

Most of us just watch the Super Bowl ads, throw around some jokes, and move on to the guacamole bowl. Others, like Marji Guyler-Alaniz, see much more and draw inspiration for food-system change. During the 2013 Super Bowl, Dodge Ram ran an ad that juxtaposed farm images with the words of Paul Harvey's now famous speech "So God Made a Farmer." While the dramatic ad quickly stirred up viral debate about the underlying message and intent,

Guyler-Alaniz clearly saw a missing piece: lack of women farmers in the ad. Rather than just vent and complain, Guyler-Alaniz took action. Building on a passion for photography, Guyler-Alaniz set out to capture images of these women, resulting in FarmHer, now an ongoing project documenting a diversity of female farmers in their everyday settings.

Debby Zygielbaum: connector of female farmers

Describing herself as the "dirt farmer and sheep wrangler" at Robert Sinskey Vineyards in California, Debby Zygielbaum lives in the state that leads the nation in organic fruit and vegetables grown by many inspiring women farmers, yet no ongoing network existed to connect women like her. Taking action to work toward filling this gap, Zygielbaum has been active in planting seeds toward a more formal network of women in sustainable agriculture in her state.

Natasha Bowens: writer

Beginning farmer Natasha Bowens does more than grow vegetables: She raises an amplified voice to celebrate the stories of Black, Latino, Indigenous, and Asian farmers and food activists working to revolutionize the food system in our communities. Moved by the fact that one doesn't see these stories in the media, Bowen started her own multi-media story-telling project in 2010, resulting in the book *The Color of Food*.

These women are a small sample of the multiple ways we are cultivating new approaches to leadership, advocating for change in our food system by cross-pollinating our strengths with our passion that keep us up at night. Much positive potential exists if we each step up and take personal responsibility to change one piece of the food system pie. It can be a small starter slice — such as inviting women in your local farmhood for that first potluck — that will undoubtedly grow to greater levels.

Remember, you don't walk this road to change alone. Your soil sisters have your back, supporting you on your way. It is in this spirit of sisterhood of the soil that I invite you to

Idea Seed:

"At some time or another, you will need to change what you are doing: Don't be afraid to go in a new direction. Change can be liberating."

— Helene Murray, executive director, Minnesota Institute for Sustainable Agriculture, St. Paul, Minnesota

continue the conversation and dive deeper into becoming a food system change agent in a free supplement to this book: *Soil Sisters: A Toolkit for Women Leaders Changing Our Food System*. In the bonus downloadable chapter, you will hear the deeper stories of the women in this book as well as a wealth of other stories and strategies to amplify your impact beyond the field.

May the pages of this book serve as starter seeds of change that both germinate your farm journey and fuel long-term commitment to transforming the health of world and planet. For me, writing these pages and interviewing the diversity of committed women farmers behind the message in *Soil Sisters* rank as a true honor. But as a writer with her heart in the field, it's time for me to power down the screen for a while and get my hands back in the soil. It's an early June morning and the strawberry harvest waits. See you down the row.

Idea Seed:

"Women: Don't back down from challenges. Stand up for your rights, and advocate for the greater good. You are stronger and wiser than you know."

— Maria Miller, director of education, National Farmers Union, Washington, DC

Tool Shed:

For additional material or resources and to keep in touch: www.soilsistersbook.com.

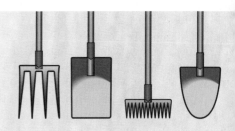

Index

About the Author

LISA KIVIRIST is a Senior Fellow, Endowed Chair in Agricultural Systems at the Minnesota Institute for Sustainable Agriculture and a national advocate for women in sustainable agriculture. She founded and directs the Rural Women's Project of the Midwest Organic and Sustainable Education Service, an award-winning initiative championing female farmers and food-based entrepreneurs.

Together with her husband, John Ivanko, Lisa is co-author of *Homemade for Sale*, *Farmstead Chef*, *ECOpreneuring*, and *Rural Renaissance*. Lisa and her family run Inn Serendipity Farm and Bed & Breakfast, completely powered by the wind and sun in the rolling green hills of southern Wisconsin.

If you have enjoyed *Soil Sisters*, you might also enjoy other

BOOKS TO BUILD A NEW SOCIETY

Our books provide positive solutions for people who want to make a difference. We specialize in:

**Food & Gardening • Resilience • Sustainable Building
Climate Change • Energy • Health & Wellness • Sustainable Living**

**Environment & Economy • Progressive Leadership • Community
Educational & Parenting Resources**

For a full list of NSP's titles, please call 1-800-567-6772 *or check out our website* at:

www.newsociety.com

3119202094645 3